NAVAL AIR STATION
JACKSONVILLE, FLORIDA

1940-2000
An Illustrated History

60 YEARS OF SERVING THE FLEET

TURNER PUBLISHING COMPANY
Paducah, Kentucky

TURNER PUBLISHING COMPANY

Publishers of America's History

412 Broadway • P.O. Box 3101

Paducah, Kentucky 42002-3101

(270) 443-0121

Written by: Ronald M. Williamson

Publishing Consultant: Herbert C. Banks II

Designer: David Hurst

Library of Congress Catalog No. 20001093925

ISBN 978-1-63026-939-5

LIMITED EDITION

Printed in the U.S.A.

Disclaimer

TABLE OF CONTENTS

FOREWORD

In mid-July 1942, with official orders in hand, I passed through the main gate of the Jacksonville Naval Air Station. I sat wide-eyed as the air station shuttle bus passed the long line of aircraft hangars. They seemed giant. Ahead were more hangars bordering the wide reach of the St. John's River. These sights remain as my first up-close memory of Naval Aviation. I was filled with awe and inspiration. Commissioned just two years earlier, Naval Air Station Jacksonville was in full operation.

With four years and a World War behind me, I returned to NAS Jacksonville. This time with orders to form, train and lead a Navy Flight Exhibition Team that would demonstrate at low altitude the fighter maneuvers so successfully used by carrier-based fighters during the war. After an intense training period, the Flight Exhibition Team flew its first public demonstration on June 15, 1946, receiving its first trophy for best performance. Shortly thereafter the team chose its name, the "Blue Angels."

Fifty years later in 1996, I again passed through a much changed main gate. It was an entrance to what now appeared to be a good size city. There were massive hangars, new buildings and facilities, and the industrial-like complex of the Naval Aviation Depot, a tenant command that repairs and reworks first line Navy aircraft for return to service to operational squadrons. This time I had been invited back to attend the air station's traditional annual air show, a nationally famous event which, this year, would feature and celebrate the 50th anniversary of the "Blue Angels."

NAS Jacksonville is now the home to 17 operational squadrons, supported by thousands of dedicated naval and civilian personnel. The station's history abounds with great presence and memories of legendary leaders in naval aviation. This book will bring to the reader a closer look into the fascinating operations and spirit of this premier Naval Air Station.

R.M. *"Butch"* Voris
Captain, U.S. Navy (Retired)
First Blue

Naval Air Station Jacksonville celebrated Sixty Years of "Service to the Fleet" on October 15, 2000. Almost 100 years have gone by since the Florida legislature passed an act designating the site for state militia training. Who at the time could have imagined what history would be made here since! Even more impressive is the fact that the site has been inhabited by early cultures starting in 6,500 B.C., some of the earliest known occupation in the United States.

Early pioneers such as A.M. Reed, or the tens of thousands of Army personnel who trained here during World War I and subsequently sailed off to war from the camp's pier on the St. Johns River, would probably recognize little at the site today. Long familiar landmarks and names have been forgotten and replaced by today's tributes to naval ships, leaders and heroes such as Yorktown Avenue, Towers Field and Jay Beasley Hangar.

One wonders what Captain Mason, the station's first Commanding Officer, might think of the place today. Some features remain as they were in 1940; however, many would probably be unrecognizable to him. One thing is certain, though. He would be proud of our Navy and its Officers, Chiefs and Sailors. Neither Captain Mason, nor any other personnel who served here in 1940, could have envisioned the achievements of NAS Jacksonville during the past 60 years.

During World War II, many pilots earned their wings at the station. Later in the war, the station was known as "the post-graduate school of Naval Aviation," teaching advanced fighter tactics. Aviators, aircrewmen and some 30,000 gunners were trained at the station to help win the war. After the war, the station remained an important part of the defense plan as it continued to play a major role in Naval aviation. So it was only natural a young LCDR "Butch" Voris would form a flight exhibition team here. Originally called the "Lancers," they are now world-famous as the "Blue Angels."

In the late 1940's and early 1950's, more than 60 percent of the Navy's aircraft on the East Coast were based here. Both the former NAS Cecil Field and Naval Station Mayport grew as satellites of NAS Jacksonville. (NAS Cecil Field, of course, has since been transferred to the City of Jacksonville as the result of Base Closure and Realignment.) The station contributed during the Korean Conflict, during the space age, the jet age, Vietnam and Desert Storm, remaining a key base for the Navy. Now a Master Antisubmarine warfare (ASW) base, it is home to the largest concentration of ASW aircraft in the Atlantic Fleet. With the recent transfer of the S-3 Viking squadrons from NAS Cecil Field, the station is now home to 17 aircraft squadrons.

This book, which has been considerably revised and updated from the 50th Anniversary edition, details the rich history of the Naval Air Station at Jacksonville and the special relationship which exists between the citizens of Jacksonville and the United States Navy. Its pages should provide something for anyone who has ever served, worked at or is just interested in Naval Air Station Jacksonville and in Jacksonville's rich Naval history.

Stephen A. Turcotte
Captain, U.S. Navy
Commanding Officer
NAS Jacksonville

ACKNOWLEDGMENTS

It has been ten years since I wrote the first book on the history of Naval Air Station Jacksonville. During the decade of the 1990's, a tremendous amount of change has taken place, both for the station and me professionally. Although still attached as the station's safety manager, I am now officially attached to Commander, Navy Region Southeast, the senior command on board NAS Jacksonville.

When the first book came out I knew I had missed important information; some information was sketchy and I would even admit there were a few errors. Although I thought I had covered a lot, I had barely grazed the surface of the history of the station. As an example, since that time I discovered a photograph of my grandfather working on the engine of an F4U Corsair at the Assembly and Repair Department in 1945. Almost from the outset of initial distribution of the first book, letters, pictures and historical memorabilia started coming in. It seemed that all a lot of the personnel who have been assigned to this great station needed to know was that someone was interested in what they had to say about their time here. To this day I still get an average of two inquiries, letters, pictures or historical items that are sent to me each week. As I write this today, a picture of a group of graduates from the Naval Air Technical Center taken in 1941 in front of the station auditorium arrived in the mail. The number of folks who have shared information about their memories of the station, as well as some of their personal momentos still astounds me.

I have been able to meet and attend the reunions of the first aviators assigned to the station. WAVES have had many reunions here and those ladies have sent extraordinary items of their WWII experiences. Personnel who were even assigned to Camp Johnston in WWI have come by to tell a story or two. I have had the opportunity to meet "Butch" Voris, the first flight leader and pilot who formed the Blue Angels, and who I was able to convince to be our special guest of honor at the station in 1994 and again for our 60th Anniversary show in November 2000. There is no finer Naval Aviator, and we keep in contact today with our weekly e-mails.

Working at the station allows me to still get to work with some of the best military and civilian personnel in the Navy and hear daily stories. John Bushick and Gregg Cunningham, leaders of the station Morale, Welfare and Recreation Department, are always good for a story or a laugh! Not to mention they have the reputation of having the finest MWR department in the Navy. Bill "Flangehead" Myers, station Airfield Manager, runs the airfield like he personally owns it. But the station could not have a finer person controlling the field and Bill again is always good for a current station story. He also always accommodates any request from me to go up in his helo for some station pictures. Contacts keep me probably one of the most informed people on the station. My wife Chris, Naval Aviation Depot safety manager, keeps me informed on the latest going's-on at the NADEP. John Kaczetow, my fellow safety brethren, keeps me up on Naval Hospital Jacksonville events, and little went on in a squadron that Larry Washburn didn't know about prior to his leaving NAS Jacksonville in 1999.

Through my historical efforts, I was able to convince many station Commanding Officers to place static display aircraft at NAS Jacksonville. This was done as a lasting tribute to all of the Naval aviators who have flown from the station and to allow the citizens of Jacksonville to see first hand some of the aircraft that have been assigned. My small collection has grown to 12 aircraft in a park-like setting, and we are looking for more.

I have to thank the current station Commanding Officer, Captain Stephen "Turkey" Turcotte, who is an avid historian. It was he who started initially pushing me to redo the history on NAS Jacksonville and gave me wide latitude to work this project. He also recently flew to the National Museum of Naval Aviation to lobby, successfully, for more additions to our static display park. I can honestly say I have never worked for a bad Commanding Officer in the 18 years I have been here. They have all been superb. Former Commanding Officer (CO) Captain Bob Whitmire collected every program from every event he attended for over 2 years just for my historical collection. No finer CO has the station ever had and why he is not still serving as an Admiral somewhere in the Navy today is a mystery to myself as well as many other station employees. Former CO Captain Skip Cramer, current director of the American Red Cross in Jacksonville, while station CO signed over 200 letters to former aviation cadets of WWII asking for their station experiences for our history files. Admiral Delaney, who was instrumental in the first history project, continues to keep in touch today on historical issues.

The Florida Times-Union was also most gracious for allowing us to again use their early station photographs. Their early articles, along with those of the former Jacksonville Journal, provided many details to the history of the Navy in Jacksonville that would otherwise be lost in time.

The personnel at NAS Jacksonville who helped in the effort for this publication were considerably larger than 10 years ago. Former Commanding Officer Bob Whitmire was most interested in assisting, as well as proof-reading. The first-hand knowledge he provided on how the Sea Control Wing actually moved from NAS Cecil Field to NAS Jacksonville could not have been obtained in any other way and I thank him for his assistance. Not enough credit was given to Captain Whitmire for his effort during this transition period! Former Commanding Officers Kevin Delaney and Charles Cramer also "chopped" their section of this book for accuracy. Lieutenant Keith Gibel wrote the section on Camp Johnston, and would have assisted further if it weren't for a pressing legal workload. Pat Dooling, Public Affairs Officer for both Commander, Navy Region Southeast and the station Commanding Officer, lent his staff to obtain some data as well as to offer any other support they could. He personally proof-read and (per my asking) made sure I didn't discover some fact that should maybe stay undiscovered! The Chief's of Staff at all three station Aircraft Wings were most helpful in obtaining squadron data. Gary Newman, station Chief of Police, provided me his collection of Scout World patches to make sure this could be included. My present work cellmate, Lieutenant Tyler Tucker, also provided proof-reading assistance as well as picture selection recommendations. The station's new Executive Officer, Captain Steven Bagby, was most excited about helping in picture selections, along with Captain Turcotte! Don Brammer, presently working on a book dealing with the squadrons stationed at NAS Miami, assisted with photographs. Also allowing me to use some of his photographs were Robert Kling, who has some 36,000 photographs mostly dealing with NAS Jacksonville and NAS Cecil Field. In addition, almost everyone at NAS Jacksonville who found out about this project, from Bill Myers offering to fly me in his helicopter for station photographs, to Larry Washburn, Dave Colburn, Mike Crivier and Phil Hoezel who would convert files and scan pictures if needed, would offer to assist in any way they could, no asking necessary! The personnel at NAS Jacksonville are truly proud of their history! Turner Publishing also is thanked for again agreeing to publish this material.

Finally, I owe the biggest thanks of all to Brian "BT" Smith. He helped with the final sections of photographs, editing, typing and proof-reading assistance, for which I would not have completed this publication by the project deadline by myself.

As in the 50th Anniversary edition, I would still encourage you to submit pictures, stories or any "corrections" to Commanding Officer, Naval Air Station Jacksonville, P.O. Box 10, Jacksonville, Florida, 32212-5000 where they will be graciously accepted. We can always straighten out that "error" in the 75th Anniversary edition!

This second volume has new information which includes detailing what squadrons were based here; what planes have been reworked at the Naval Aviation Depot; and gives a chronology of significant station events so it is easier for the reader to research a particular area of interest. Information has been expanded considerably from the first printing, and even a few errors straightened out! Does this book tell the whole story of NAS Jacksonville? It doesn't even come close! But it does give the reader an idea of how this station came to be and where it might be heading in the future. I hope you enjoy reading it as much as I again enjoyed writing it!

Lastly, I must again thank my wife Chris who not only provided support but also let me keep my NAS Jacksonville stuff scattered throughout a room of our house as I collated for this publication!

Ronald M. Williamson

FIGURE 1.
PROJECTILE POINTS FROM NAVAL AIR STATION JACKSONVILLE SITE

A. Newnan point (ca. 5000-3000 B.C.)
B. Culbreath point (ca. 3000-2000 B.C.)
C. Levy point (ca. 5000-1000 B.C.)
D. Archaic stemmed point (ca. 5000-1000 B.C.)

FIGURE 2.
POTTERY SHERDS FROM NAVAL AIR STATION JACKSONVILLE SITE

A. Orange Incised (ca. 1650-1450 B.C.)
B. Orange Incised (ca. 1650-1000 B.C.)
C. Deptford Cross Simple Stamped (ca. 500 B.C.-A.D. 700)
D. St. Johns Check Stamped (ca. 750-1565)
E. Savannah Cord Marked (ca. A.D. 1200-1450)
F. Savannah Cord Marked (ca. A.D. 1200-1450)

Photos courtesy of Florida Archeological Services, Inc., Jacksonville, Florida

"Near this majestic oak, Pedro Menendez, founder of St. Augustine, parleyed with Ucita, Chief of the Timquans. Here Oglethrope established headquarters during his campaign against the Spaniards at Fort Picolata up the Saint John's River. In 1812, from the shadows of this oak, American forces aided by the President of the Republic of Florida, who owned the surrounding acres, launched an attack against the Spanish on the east bank of the river." These words are on a plaque now standing in front of a small oak tree in front of the NAS Jacksonville Safety Office. For many years the plaque stood in front of a massive Timucan oak tree on Mustin road in the officers' housing area. The information on the plaque was used from 1942 to 1965 in practically every article written about the early history of the area now occupied by NAS Jacksonville.

In 1960, when the air field was finally dedicated as "Towers Field", this information was used in the official program. The information, however, cannot be historically documented to Naval Air Station Jacksonville grounds nor the oak tree site. It seems that Rear Admiral Gilchrist B. Stockton, a prominent Jacksonville-born resident present at NAS Jacksonville in 1942 as a Commander in Charge of Public Relations, paid for the plaque and monument out of his own pocket, made up the inscription, and then placed the monument in front of the oak tree.

Why? As the base was being built in late 1941, the oak tree was scheduled for removal. Commander Stockton, who was also instrumental in helping to establish the Naval Air Station in the late 1930s, evidently devised the historic scheme to save the magnificent tree. It worked almost too well—as the numerous articles written using his monument's information attest to. The monument was not removed until 1978, under pressure from the Jacksonville Historical Society. Today, a correct historical marker is in front of the massive oak tree on Mustin Road. That Timucan oak tree was barely beat out by an oak tree in Gainesville as the state's oldest champion in 1983.

The actual early history of the land occupied by NAS Jacksonville is not quite as colorful as the monument

indicated, and is now known to have a history that is as old as any in the United States. In 1995, an archeological study was conducted at the station. This study, conducted by Robert Johnson, has provided a wealth of new information concerning the early history of the site now occupied by NAS Jacksonville. The following section, written by Robert Johnson, provides a brief overview of the real early years history!

BEFORE EUROPEAN

CONTACT

by
Robert E. Johnson, archeologist
Florida Archeological Services, Inc.

Before the lands now known as NAS Jacksonville were settled, numerous game animals roamed the nearby forests and woodlands. In addition, the Mulberry Cove area of NAS Jacksonville at Black Point, and the St. Johns River contained bounties of fish and other resources. These facts were well known by the earliest inhabitants of the area. Known as the Early Archaic culture, peoples of this period were nomadic hunters and gatherers who occupied short-term camps on the shores of the St. Johns at what is now NAS Jacksonville as early as 6500 B.C. These early peoples left evidence of their utilization of the lands in the form of stone tools, known as projectile points or lithic artifacts to archeologists. Such tools were discovered during an archeological survey of NAS Jacksonville in 1995 and 1996 (conducted by Florida Archeological Services, Jacksonville), a survey that yielded an impressive 36 archeological sites (Johnson 1997). These points have been given names such as Arredondo and Greenbier, and were found at two archeological sites at NAS Jacksonville.

During the subsequent Middle Archaic period, archeological deposits of the area contained variants of the Florida Archaic Stemmed point which were utilized by local aboriginal populations during the period from 5000-3000 B.C. Projectile points such as Newnan, Levy, and Marion have been found by archeologists at sites of this

period (Figure 1). Radiocarbon dating techniques have been used to determine that the coastal area of Duval County appears to have supported sustained occupations by aboriginal populations of this period around 3700-2300 B.C. (Russo 1992). Further south along the St. Johns River below the community of Palatka, fresh water shell middens represent the residue of aboriginal occupation in which the peoples of this time period, also known as the Mount Taylor culture, ate animals from the forest, as well as mystery snails and river mussels, the shells of which are abundant in archeological sites of the period. While such deposits have been found in the vicinity of the Navy's Rodman Range in Putnam County, none were discovered during the NAS Jacksonville survey.

It was during the subsequent Late Archaic period that one of the most intense early aboriginal occupations of NAS Jacksonville occurred. Of 36 archeological sites identified in the environs of the base, 22 of these contained evidence of this period indicating that the base sustained relatively intensive use during the period from 2000 B.C. to 500 B.C. (Johnson 1997). It was during this period that one of the most important technological innovations occurred in northeast Florida. This era marks the first use of aboriginal populations of the region of fired-clay pottery. Known by archeologists as Orange pottery, this pottery type, which is the oldest ceramic ware in North America, is believed to have been tempered with locally available vegetable fibers such as Spanish moss. The NAS Jacksonville archeological survey recovered thousands upon thousands of ceramic artifacts of this and subsequent cultural periods. During a similar archeological investigation at Kings Bay, Georgia in 1978, a Late Archaic period fire pit (or hearth) yielded incised fiber-tempered pottery of the Orange period and a series of radiocarbon dates ranging from 3050-2310 B.C. (Johnson 1978).

Elsewhere in Duval County, Late Archaic period sites proliferate indicating an increase in local aboriginal populations. Along the coast, large sites of the period contain shellfish remains such as oyster and clam. At such sites, archeologists have utilized field techniques which have led to the reconstruction of aboriginal dietary patterns and the documentation of coastal aboriginal subsistence strategies. At NAS Jacksonville, little or no shellfish remains were found indicating that these peoples probably came to the area to exploit the abundant floral and faunal (plant and animal) resources of the nearby forest. In essence, peoples of the Late Archaic period are known as fisher-folk and shellfish collectors, with terrestrial mammals and locally available plants such as hickory nuts, acorns, and berries also contributing to the diet. While little is known of the structures built by these Indians, they are believed to have adapted quite well to the coastal/riverine environments of the region. At this time period, the forests of NAS Jacksonville would have contained abundant resources facilitating such adaptation. While the forests contained nuts and berries, animals such as white-tailed deer, bear, and turkeys would have also been present and exploited by aboriginal populations.

In the millennium before European contact (500 B.C. to A.D. 1565), the NAS Jacksonville area was occupied by Indian populations of the St. Johns culture among others. Material culture collections (i.e. artifacts) of this period are generally represented by mixed pottery assemblages containing non-diagnostic plainwares which have no distinguishing surface treatment that can enable placement of a particular occupation within a certain time period. In contrast, other pottery wares of the period contain diagnostic temper (i.e. clay additives) and surface treatment (i.e. decorations) which allow archeologists to further divide the period. During the initial phase of this two stage era known as St. Johns I, or from 500 B.C. to A.D. 750, ceramic wares known as St. Johns, Deptford, and Swift Creek, each with distinguishing paste and surface decorations, are commonly found in numerous sites in the region (Johnson 1988). However, only a few culturally pure sites have been discovered in Duval County, two of these in close proximity to Mayport Naval Station near the mouth of the St. Johns River (Johnson 1998a; 1998b). Recent archeological investigations in the region have suggested that these mixed assemblage sites should constitute their own regionally important cultural recognition. The term "St. Marys" culture, has been posited for the northeast Florida area (Russo 1992). Additional archeological research is needed, however, before this scientific view can be substantiated.

During the period from 500 B.C. to A.D. 1565, aboriginal populations of the region increased in size and cultural complexity. There was an increase in the exploitation of coastal resources, new pottery styles appeared, and the Indians began to expand their religious beliefs by placing their dead in burial mounds. While it has been suggested that there was an increase in plant cultivation during this time period, little firm archeological evidence has been discovered to indicate that horticultural practices were widespread in the region.

Around A.D. 750-800, St. Johns Checked Stamped pottery was first utilized at NAS Jacksonville and other sites in the region. Marking the onset of the second stage of the period (or St. Johns II), it was wares of this type, as well as those of the St. Johns I period, that were expected to be the most prolific in the artifact collections of the NAS Jacksonville archeological survey. In fact, nearly half of the sites identified at NAS Jacksonville contained St. Johns ceramics. Other Indian pottery types common to the region include styles called Deptford and Swift Creek. While Swift Creek pottery was virtually absent from archeological sites at NAS Jacksonville, Deptford ware pottery was recovered from 11 sites at the base.

While these archeological data indicate that NAS Jacksonville was utilized by native aboriginal populations to some extent from 500 B.C. to A.D. 1000, these occupations were not as extensive as those during the previous Orange period (2000 B.C. to 500 B.C.), nor those that followed post-A.D. 800 times. It was during this latter period, or sometime between A.D. 800-1000 and A.D. 1565 at European contact, that the most extensive aboriginal occupations occurred at the base. Archeologists were surprised to learn that the most extensive Indian sites at NAS Jacksonville were those of the Savannah culture, a culture whose heartland lies in the coastal Georgia area near present-day Savannah. Prior to the archeological studies at NAS Jacksonville, it was anticipated that the most abundant Indian pottery would be those of the St. Johns cultures, but by far the most extensive pottery types belong to the Savannah culture (Johnson 1997).

Although there is still some debate among archeologists concerning whether Savannah Cord Marked and other pottery types attributable to this culture were manufactured locally or entered the area by trade, such wares were recovered from 18 of the archeological sites identified at the base. Moreover, the largest site discovered at NAS Jacksonville dates to this period. Covering some 40 acres in extent, this site yielded nearly 9,500 artifacts. The majority of these artifacts are believed to be of the Savannah culture. The lack of European (i.e. Spanish) artifacts in this site indicates a pre-contact occupation. Moreover, the lack of pottery made by the local Timucuan Indians (known as San Pedro wares), as well as those of the Guale Indians of the post-contact Georgia coast (known as San Marcos wares), suggests that the Savannah inhabitants at NAS Jacksonville were either not in contact with these peoples, or more likely actually preceded these later occupations. This information suggest that the NAS Jacksonville Savannah period sites probably date to the earlier period of Savannah occupation in the region rather than during the latter.

In addition to pottery and stone tools, evidence of aboriginal structures was also found in one Savannah period site suggesting the possibility of relatively long-term occupation. Archeological evidence indicates that the Savannah peoples that occupied NAS Jacksonville likely returned repeatedly over perhaps a 300-500 year period to the same locale to exploit the abundant natural resources of the nearby forest and river. Year-round occupation is a possibility but has not been documented. Radiocarbon dates from Savannah sites in the northeast Florida region suggest an occupational range from A.D. 1180 to A.D. 1520. Averaging about A.D. 1355, these dates correspond well to the anticipated Savannah period occupation at NAS Jacksonville.

Although some limited archeological evidence was found during the NAS Jacksonville archeological study to suggest that the base was occupied during the early historic period, no firm evidence of Mulberry Plantation, or any other historic period site, was discovered during the study. By this, the historic period, all aboriginal occupations at the base had terminated. As such, there appears to have been little or no Native American Indian presence at NAS Jacksonville at the time of European contact.

European Contact

Three Spanish grants accounted for most of the early European history.

The first grant was Pointa Negra, Spanish for Black Point, given to Don Felipe Bastros. The second grant was Sans Souci, granted to A. C. Ferguson and the third grant was Moral Grueso (meaning Fine Mulberries) belonging to T. Hollingsworth. A. M. Reed eventually bought most of the property which became known as Mulberry Grove Plantation. The one time plantation area is now occupied by the officer's housing area.

Between 1763 and 1784, when the British occupied Florida, the site was first changed from a wilderness area to a more hospitable place for civilized man. Three plantations were established along the river and crops of cotton, corn and sugar were raised. When the Spanish regained control of Florida in 1784, San Nicholas, an old abandoned fort, was rebuilt as headquarters at Jacksonville and the Spanish governor started levying heavy taxes. John Houston McIntosh, a planter at Ortega Point, grew tired of paying the heavy taxes and started stirring up individuals to acquire the territory in the name of the United States. In 1812, McIntosh became President of the short-lived Republic of Florida. Shortly thereafter, he acquired the three grants which included most of the acreage NAS Jacksonville now occupies. In 1821 the Florida territories finally became a possession of the United States. Plantation activities declined, and the area was again becoming a woodland.

In 1860, as the War Between the States was heating up, cotton became an important crop to the southern cause. Fields of white bolls appeared all across the site. This continued until the late 1880s when the local economy started to decline. The future base site again started returning to wilderness. A. M. Reed, his wife and two daughters moved to the site in 1862 and lived in a large house overlooking the St. John's River in the approximate area of Quarters H of today. There is an old story concerning a cache of buried silver aboard the NAS Jacksonville site. The War Between the States brought fear into the Reed Plantation family, and they supposedly buried large amounts of silver in the areas occupied today by Officers Housing, and west of that area

into and including the current Weapons Compound-Golf Course area. George Person, A. M. Reed's great-grandson, stated "During the war, grandfather Reed, fearful of raiding federals, buried a great deal of the family silver on the place. Grandfather Reed was the only one who knew exactly where and when he died, the secret died with him."

Later accounts state that approximately 60 percent of the silver was eventually found. The rest is for the finding! In 1939, when the Persons made their deal to sell their land for use by the Navy, an agreement was reached that if the silver was uncovered by excavation, it would be turned over to the Persons if they were able to properly identify the pieces of silver found. Since 1940 no silver has yet been found.

Mr. Reed kept a diary from 1874 to 1892, a copy of which is located at the Florida Times-Union library. The typical entry (per a Florida Times-Union article) for February 17, 1874 read "Hands hunting cattle - clearing berry patch - building goose pen - putting fence around poultry yard. Guests, as usual, at Mulberry Grove." The diary also contained some humorous entries. On March 16, 1876, Reed wrote "All hands gone to Harry Flowers wedding (but the bridegroom did not come.)" Another wedding by plantation hands prompted the diary entries in May 1881. The May 3 entry said "Jack arrested for breach of promise." and then on May 12, "Jack married."

The diary also listed news events such as the earthquake of January 13, 1878. In 1883, Reed signed a lease giving the Atlantic Coast Line Railroad right of way. (The same track exists today west of U.S. 17). On March 7, 1884, the first passenger train stopped at Reed's landing. Freight was delivered for the first time on January 1, 1885. By February Mr. Reed was complaining the railroad was killing his cattle. Prior to the railroad, steamboats were the primary method for receiving guests. A. M. Reed died in 1889 at age 86.

In 1905 family members began selling off parcels of Mulberry Grove Plantation and the last 336 acres of the original 1400 acre plantation was sold for establishment of NAS Jacksonville in 1939.

Much of the future NAS Jacksonville site was quickly being turned into subdivisions by 1906 with Philbrofen subdivision platted on March 6, 1906. This area included all of the area

east of present day Mustin Road to the St. John's River. Present day Wright Street was known as Philbrofen Avenue and Mustin Avenue was known as Riverside Drive.

On the recommendation of Adjutant General J. C. R. Foster an act was passed by the Florida legislature in 1905 establishing a commission to select a permanent site for state militia training.

After visiting numerous sites, the commission selected the area near Yukon in 1907. The first tract acquired, the Philbrofen subdivision, consisted of 300 acres. As additional land was for sale, the War Department devised a plan to obtain it for rifle practice. The commission reported it held an option to buy the property for $20 an acre.

The citizens of Jacksonville had already raised $6000 to establish the camp, and $8000 in Federal money was made available to purchase additional properties. Thus, purchases were made by the Federal Government as follows: August 10, 1907, 400 acres; September 11, 1908, two tracts, one for 85 acres and the other for 108 1/3 acres and on November 18, 1913, 100 acres. The State of Florida's 300 acres, combined with the Federal Government's 693 1/3 acres, for a total of 993 1/3 acres, saw the beginnings of a permanent military presence.

The state approved the commission's recommendations and the camp site was on the way to being a reality. The first encampment of state troops at Black Point was June 8-15, 1909. Rifle competitions were held in 1915 and 1916 at the second largest rifle range in the U.S. Combined funds spent on the site for improvements by Federal and state Government totaled $250,000 by 1917.

On December 4, 1916, Earl Dodge, a New York millionaire, opened an aviation training camp at Black Point. When World War I was declared, aircraft from that training camp flew over Jacksonville with the most advanced Curtiss aeroplanes of the day flew in and out of the airfield. The aviation training camp operated for about 8 months and not a single fatal accident occurred during flight operations. On April 18, 1919, Maj. T. C. McCauley landed at Camp Johnston after completing a cross-continental flight from San Diego, California, in 25 hours and 45 minutes flying time.

SOUVENIR ~

of

Camp Joseph
E. Johnston

Jacksonville, Fla.

Camp Johnston Park

The Greatest Values of Improved Country Estates Ever Offered

Adjacent To Jacksonville.

Extra Large Estates From 1-4 to 3 Acres Each

Only **$200 and up**

15 Per Cent Cash—Balance in 60 Equal Monthly Payments With 7 Per Cent Interest On Unpaid Balance.
IMPROVEMENTS ALONE WORTH MORE THAN PRICE ASKED FOR LAND.

RIVERSIDE AVE. 1902.

WHAT PEOPLE ARE SAYING.

Under Date of July 22, 1922,
Roger W. Babson, The Statistical
Expert of America said:

My advice to those who want
suburban real estate is to buy at
once. For the past decade or more,
people have been crowding to the
cities. Now, however a reverse
movement to the country is beginning which promises to be the

U. S. GOVERNMENT SPENT

Hundreds of Thousands of Dollars
Installing the Present Improvements.

THEIR LOSS YOUR GAIN.

Camp Johnston Park Estates are

In September 1917, the Department of War took over the reservation for war purposes, and leased additional lands from private owners, naming the site Camp Joseph E. Johnston. In General Joseph E. Johnston, C.S.A., A Different Valor, by Gilbert E. Govan and James W. Livingood, mention is made of General Johnston's assignment by President Cleveland in July 1885 to act as one of the pallbearers at Grant's funeral: "It was a fitting assignment as each man held great respect for the other. Grant had singled Johnston out as the ablest of the Confederate commanders and the one who had given him the most anxiety. Shortly before his death Grant had written of the campaign in Georgia that neither the great defensive soldier, Thomas, nor the ingenious Sherman, 'nor any other soldier could have done it better' than Old Joe." Govan and Livingood write, "Ironically, much of his [General Johnston's] recognition has come to him because of the campaign in which his commander in chief most doubted him, that against Sherman from Dalton to Atlanta. Although overshadowed by the postwar Davis legend and overlooked by biographers since 1893, Johnston is accepted as one of America's great commanders." Notwithstanding the above, Camp Johnston was most likely named after General Joseph E. Johnston, because General Johnston was the Quartermaster General of the United States Army before tendering his resignation to join the confederacy on April 22, 1861.

Camp Johnston was not established without a political battle, however. As soon as World War I was declared, Federal representatives in Congress had been contacted by citizens in Jacksonville asking for the establishment of a cantonment at the Black Point site. General Leonard Wood, who was to pick sites for sixteen cantonments, sent an aide to inspect the Black Point area. The day he came was rainy, and the aide left with a most unfavorable recommendation. The three problems cited in his report included defective terrain, inadequate water supply, and the area being mosquito-ridden and malarious.

Mr. W.R. Carter, editor of the newspaper "Jacksonville Metropolis," decided he would fight for establishment of the camp. His efforts were responsible for General Wood making a personal trip to Black Point on June 25, 1917. He was so impressed he returned to Washington and recommended Jacksonville as a site. But the good news did not last long. The Army changed the site from Jacksonville to another location due to military reasons. Soon, however, the Army had to select another site, this time for a Quartermaster Training Camp. Washington, D.C., seemed to be the prime site for that camp, but questions arose about having the camp as far inland as Washington, and the selection was at a stalemate between Washington and Jacksonville.

With the matter swinging in the balance, General Francis J. Kernan, a West Point graduate and a Jacksonville native, was asked for his views. He recommended Jacksonville and the recommendation was accepted in early September 1917. One condition of selection was that liquor be kept away from the soldiers to be stationed there.

On September 10, 1917, Professor J. S. Pray, Professor of Landscape Architecture at Harvard University, was hired by the War Department to visit the site and lay the camp out. One month later, after some arguing with contractors, actual construction work began on the $2.9 million project. On September 24, 1917, the Officer in Charge of Cantonment Construction, I.W. Littell, Colonel, Quartermaster Corps, "by the authority of the Secretary of War," directed the Constructing Quartermaster of the Quartermaster Training Camp in Jacksonville, Florida, Major F.I. Wheeler, Engineer Reserve Corps, to construct the Quartermaster's Training Camp near Jacksonville, Florida. The letter stated that the contractor for this work would be A. Bentley & Sons, of Toledo, Ohio and the Supervising Engineer would be Lockwood Green & Company of Atlanta, Georgia. On October 13, 1917, Lt. Col. Fred L. Munson and his staff arrived, and a few days later, they assumed formal charge of Camp Joseph E. Johnston.

The general contractors, A. Bentley and Sons Company, along with supervising engineers Lockwood, Greene and Company and Wilson Construction Company (road contractors) constructed the entire camp in just under four months. At the peak of construction activity, almost 9,000 workers were busy at the camp. Even with this tremendous construction activity, only one worker was killled.

In Major Wheeler's completion report, dated April 22, 1918, he made various observations as to the site, climate, landscape, attitude of local people, influence on business, clearing and drainage, roads, organization of construction forces, administrative employees, transportation, housing and feeding laborers, policing, Sunday work, schedule of wages paid for labor, commissions to foreman, accidents, preparation of report, and employment on the Army on construction work in the United States. Most of the findings in his completion report are reprinted below:

"1. SITE.

Camp Joseph E. Johnston is located at the Florida National Guard Camp Ground at Black Point on the west band of the St. Johns River about eleven miles above the City of Jacksonville.

The selection of this site was in many respects fortunate; lumber was available for camp construction in sufficient quantities by rail or barge on short notice; the climate is such as to allow construction to proceed all winter without interruption; the proximity of a large river provides for the economical disposal of sewage; the fine sandy soil furnishes agreeable footing even in rainy weather; the construction of serviceable roads is not a difficult problem.

3. LANDSCAPE.

The site selected for the camp possessed many points of natural utility and beauty and in locating the new work these were preserved as far as practicable. There were many beautiful live oaks and magnolia trees upon the reservation, some of great size, and these were cut only when absolutely necessary to obtain space for buildings and roads. Many buildings were shifted out of the typical locations in order to leave in place the large or beautiful trees. This was particularly true at the Base Hospital site, where the group of buildings was first roughly staked and then shifted to preserve a maximum of the desirable natural features. Several marshy areas, the devation of which was but little if any above tide water, were filled to prevent the breeding of mosquitoes.

4. ATTITUDE OF LOCAL PEOPLE.

The people of Jacksonville were very hospitable and endeavored to assist in every way possible in advancing the construction of the camp. It is believed they were actuated by motives both patriotic and practical.

5. INFLUENCE ON BUSINESS.

It is a well known fact that business in Jacksonville, especially the lumber industry, was practically dead at the time Black Point was selected as the site of the Quartermaster Training Camp. Due to purchases of large quantities of material in the vicinity, the employment of large numbers of men at the camp and the incidental influx of a great many people, business in Jacksonville picked up very rapidly . . .

7. ROADS.

The site had been used as a State Military camp and was provided with a brick road from the County Highway entirely through the camp grounds. This road was 16 feet wide and of brick laid flat on sand foundation with a concrete curb . . . With only minor repairs this road carried the entire traffic of the camp for four months, including the construction period. Much of the road between Jacksonville and camp was of the same type and only 16 feet wide. A road of this width was, of course, too narrow for safety. It was necessary at the time the largest number of laborers were employed (about 8400) to restrict traffic to one direction at the time the employees were going to, or from, camp . . .

[The remainder of the Major's observations under this heading are from a letter he sent to the Officer in Charge of Cantonment Construction, Washington, D.C. on Feb. 16, 1918]

. . . It is my understanding that the City of Jacksonville made certain promises that induced the War Department to select Black Point as the site for the Quartermaster Training Camp. The promise to extend the suburban electric line to Camp Johnston is now being fulfilled and it is expected that the line will be in operation in March. The City has provided one additional artesian well in accordance with promise, and the camp is supplied with abundance of pure water. The promise to make the City morally clean is now being fulfilled under pressure of the Mayor, the Commander and other government agencies. The local representatives also guaranteed that a satisfactory highway from Jacksonville to the camp would be constructed but thus far no progress toward this improvement has been made . . .

. . . The road has stood up beyond expectation under increasing traffic but is now getting in bad condition and unless given constant care will soon become dangerous . . .

. . . The business interests of the City have benefited greatly by the location of the camp in its vicinity and are under obligation to fulfill their promise and provide a roadway commensurate with the traffic incident to the camp activities. It is suggested that the War Department remind the City of its guarantee and obligation.

8. ORGANIZATION OF CONSTRUCTION FORCES.

The Constructing Quartermaster and his Commissioned Assistants, the Auditor, the General Contractor, Supervising Engineers and Road Contractors were selected by the Officer in Charge of Cantonment Construction.
The General Contractors, A. Bentley & Sons Company, had completed Camp Sherman; the Supervising Engineers, Lockwood, Greene & Company, held the same position at Camp Gordan; the Road Contractors, the Wilson Construction Company, had constructed many miles of paved highways in Florida. The reputation for

efficiency of all these firms was therefore well established and the Constructing Quartermaster therefore placed great confidence in their judgment as to handling the work . . .

9. ADMINISTRATIVE EMPLOYEES.

There follows a list of the administrative forces in November and December, the number then employed being 6,000 to 8,400. The rates paid by the contractor appear somewhat high, but it must be remembered that at that time the demand for the class of service required was excessive and no promise could be given as to the length of period of service. Efficient employees had no difficulty in securing employment and could not be induced to leave their positions except by an offer of a high rate of compensation. The economical handling of a large number of laborers depended upon the presence of competent supervisors. The stipulation was made by the Secretary of War that the camp was to be constructed in the shortest possible time,—economy was an important consideration but secondary to speed of construction. Under the circumstances it is believed that the expense of supervision was warranted . . .

16. LABOR.

It was necessary to bring laborers from surrounding communities and pay railroad fare to the camp in order to get sufficient men of desirable character. The labor employed was, as a rule, efficient but there appeared in the ranks many who were mediocre or entirely worthless and these were eliminated as fast as discovered by foremen or labor inspectors . . . On several occasions the laborers requested audiences with the Constructing Quartermaster for the purpose of securing higher rates of wages or to present grievances and several conferences were held with officials of labor organizations. On two occasions rates of pay were increased in consideration of the length of time required to travel between Jacksonville and camp and the rates in force at other camps. A request for recognition of the Closed Shop principle by the Carpenters Union was denied and attention called to the explicit direction of the War Department that the employment of labor in the construction of camps must be absolutely on the Open Shop basis. The strike threatened by the Carpenters Union did not materialize.

17. TRANSPORTATION.

The problem of conveying employees between Jacksonville and camp was a difficult one to solve. A steamer with a capacity of about 1100 person was chartered for $200.00 per day. This was a very economical means of transportation but only a part of the employees could be carried in this way. A request was made to the A.C.L.R.C. to furnish a shuttle train between Jacksonville and the camp but they were able to provide only cattle and box cars, upon which the charge was $37.50 per car per day and the capacity of the cars was limited to 75 persons. These cars were unlighted although the laborers were obliged to travel before light in the morning and after dark at night. The A.C.L.R.R. claimed entire inability to provide passenger coaches on account of Government requirements for troop shipments. Only common labor could be transported in this way as high class laborers refused to ride in these cars. In order to hold the higher class of skilled labor it was necessary to provide automobile transportation. Several touring cars and trucks were employed to convey Superintendents, foremen, time-keepers and office employees. These cars and trucks were available

also for transportation over the camp during the day and thus proved to be an economical means of transportation. Some of the employees were carried in busses at 35¢ to 50¢ per trip. The rate later established by the Camp Commander for automobile transportation between Jacksonville and Camp was 50¢ per trip. It is believed the A.C.L.R.R. should have been required to provide transportation by suitable shuttle trains at 10¢ or 15¢ per trip.

19. POLICING.

There was little conflict of authority between the construction forces and the permanent military population of camp. I attribute freedom from conflict of authority largely to the employment of a civilian police force to take care of the policing of the construction forces and property. The military police attended to the discipline of the troops. It has been noted that there is often a lack of sympathy and cooperation between civilian and military forces, especially where the Military is other than that pertaining to the Construction Department. The Construction forces are interested primarily in the progress of the work and only secondarily in refinement of Military formalities, while the military authority is often more interested in the details of military procedure.

21. SCHEDULE OF WAGES PAID FOR LABOR.

CAMP JOSEPH E. JOHNSTON, FLA.

CLASS OF LABOR	RATE OF PAY PER HOUR
Blacksmith	.55
" helper	.40
Bricklayer	.75
" foreman	.80
" helper	.40
Carpenter	.55
" foreman	.70
" helper	.40
Cement Finisher	.75
Cement Mixer	.30
Cement Worker	.30
Chauffeurs	.25
Electricians Outside Foremen	.65
Electricians Groundman	.30
Electrical Helper	.40
Electrician Laborer	.25
Electrician Lineman	.60
Electrician Lineman Helper	.45
Electric Wire Foreman Inside	.75
Electric Wiremen Inside	.621/2
Engineer, Gas	.65
Engineer, Ditch Machine	.65
Engineer, Locomotive	
Engineer, Stationary	.65
Engineer, Steam Roller	.65
Fireman, Locomotive Glazier	.55
Labor, common	.20
Labor, Foreman	.60

NOTE: Laborers were paid time and one-half for all time in excess of 8 hours and double time for Sunday and Holidays.

22.   COMMISSIONS TO FOREMEN.

It was discovered that some of the foremen were accepting presents from the laborers, theoretically in recognition of appreciation of their ability and agreeable treatment of the workmen. In some cases there was very strong evidence of the laborers paying money to hold their jobs. Whether this was true or not it was regarded as a bad custom, destructive of efficiency and was prohibited. Wherever there was strong evidence of the foreman receiving a commission for retaining an employee, the foreman and employee were both discharged. Evidence of such transactions was obtained with great difficulty on account of the reluctance

CLERICAL EMPLOYEES OF CONSTRUCTING QUARTERMASTER'S OFFICE		
NAME	*POSITION*	*RATE OF PAY PER MONTH*
VAN HOUTEN, W.E.	Chief Clerk	$ 200.
RITCHIE, Jos. S.	Cost Clerk	135.
WALSH, J. J.	Stenographer	125.
WAGNER, H. M.	Cost Clerk	110.
LITTLE, H. A.	Stenographer	100.
PARKER, John	"	100.
GILMER, J. T.	File Clerk	100.
FARLEY, George	"	75.
REYNOLDS, S.G.	Messenger	45.
STALEY, John R.	"	40.
GOVERNMENT AUDITING STAFF		
MUCKLOW, Walter	Auditor	$ 350.
BENNETT, Russel W.	"	350.
GREELEY, Allan	Chief Time Inspector	300.
WOODROW, D.S.	Chief Clerk	250.
MERRY, Norman	Chief Voucher Clerk	200.

on the part of laborers to giving testimony. Inefficient and lazy workmen are willing to share their pay with a foreman for the sake of remaining in a gang of skilled laborers where they can have agreeable employment and a higher rate of pay than they are capable of earning. It is believed that every large body of laborers should have assigned to it a few secret service men, in close touch with the laborers, without being known as such. It is only in this way that collusion can be detected.

23. ACCIDENTS.

. . . A laborer in alighting from a work train before it stopped, fell beneath the car and was instantly killed. The laborer was at fault.

24. PREPARATION OF REPORT.

In order to make this report as valuable as possible and to secure statements from several points of view the Constructing Quartermaster requested the co-operation of his assistants, agreeing to give credit for matter submitted. Some of the heads of departments responded and their reports are appended.

25. EMPLOYMENT OF THE ARMY ON CONSTRUCTION WORK IN THE UNITED STATES.

It is understood that nearly all construction work in France is now being performed by the Army. In the United States practically all camp construction is performed by civilians. The employment of civilian labor involves frequent controversies as to rates of pay and hours of work. It is found impracticable for the civilian employees to provide housing and feeding facilities and it is necessary for the government to provide these accomodations. It frequently happens that when a new job starts laborers will be drawn from other localities where they are needed and it is impossible to prevent this.

Under these circumstances it is believed advisable to employ enlisted labor for camp construction. This would not be a departure from present practice. A vast amount of work in camp is performed by enlisted labor. This is done largely by the Camp Quartermaster's detachment which is practically a permanent camp institution. The Camp Quartmaster's detachment could easily be expanded to

perform additional work required in connection with camp construction under the Constructing Quartermaster. These men should however be entirely disconnected from the fighting units and not required to report at any time for drill, target practice or any other duty as irregular service of men on the construction work would be detrimental. It is believed entirely practicable, in enlisting men for this special service, to waive many physical defects which would cause rejection when enlisting men for the fighting units. I see no objection to having men permanently assigned to the various mechanical trades and I believe this scheme could be worked out to advantage in the construction of camps. The great advantage in this arrangement is in the fact that the housing, feeding and transportation problems are already solved as the men are permanently quartered and fed at the camp. It might be necessary to employ civilians for some of the positions but I believe that the scheme of using the Army for construction work in this country could be as successfully worked out as in now being done in France.

During the construction of Camp Johnston, men of all grades and classes have been employed, some of them at very high wages, to perform duties which might have been performed by commissioned and enlisted personnel at

Camp Johnston—wide angle views.

less cost. I realize that work of such magnitude as the construction of camps requires very high grade men but I believe under present circumstances it is entirely feasible to find such grade of men in the Army by offering them commissions, or by enlisting them. Mr. Bentley has informed me that many of the men in his organization would be willing to accept commissions and perform the same duties which they are now performing. Such an arrangement would be advantageous for the reason that there would be less duplication of service . . ."

It is interesting to note Major Wheeler's observations of the City of Jacksonville and its relationship to the military. Also of interest are the Major's thoughts on the use of contractors versus military to perform various tasks. Many of the issues the Major addressed in his completion report are still being discussed today.

H. Davis, in History of Jacksonville, Florida, *detailed the arrival of the first soldier at the camp.*

"On the morning of October 16, 1917, a soldier in uniform appeared at the office of the contractor at Camp Johnston and said he had arrived to report for duty. The contractor was astonished and did not know what to tell him as work on the camp has just started, and it was not ready to receive recruits. Colonel Munson was hunted up and it was decided to provide quarters for the lone soldier and fix him up a mess. So Private Barclay, sent here by mistake from Camp Custer at Battle Creek, was the first private soldier to enter Camp Johnston."

The first group of officers and enlisted men arrived for training November 19, 1917, and by December 3, 1917, they had left for battlefronts in Europe. In December, the camp was selected as a remount station, Auxiliary Remount Depot 333, and 160 acres (of which only 103 acres were actually cleared) near Yukon, FL were allotted for its use. The depot consisted of 16 buildings, 14 stables, and had provisions for 4,000 horses and mules.

Old tales from Army men who trained at Camp Johnston indicated they experienced difficult times.

Mr. William H. Hottinger recalled the following information about the camp: "I remember alligators in and around the camp as thick as fleas. They were 6 and 10 feet long, and it seemed that more ammunition was expended in

shooting the alligators than on the range. Snakes, too, great big ones, were bothersome. There was a terrific death rate from the flu, meningitis, and certain types of malaria."

Mr. Bill Potts recalled, "The camp was hot, and that sand was ridiculous to try to march in. The heat and rain were nearly overpowering."

Notwithstanding the above anecdotes and the initial unfavorable report by General Wood's aide, Camp Johnston proved to be one of the healthiest camps, with a pro rata sick and death rate as low as any camp in the country. The only problem was the excessive drunkenness among soldiers at the camp. The situation, however, was rectified by Duval County voting "dry."

The small former National Guard camp quickly expanded to several square miles with roads built with tons of fill, then sand and brick on top. Most of the barracks were built on stilts and the ground was filled in around them because of the swampland. The population of the camp peaked at 27,000 men, most of them departing from Black Point after a brief training period, for service in Europe.

Most of the recruits, all who were designated Quartermasters regardless of qualification, arrived at Camp Johnston by train. The trip from Chicago to Jacksonville averaged six days. Local transportation was primarily by ferry or rear wheel paddle steamer. The ferry crossed the river to Mandarin, and for 15 cents, a soldier could ride the steamer to Palatka. A steamer trip to the mouth of the St. John's River cost 75 cents.

Life at the camp did have a touch of the Old South elegance. Officers dining in the Headquarters Mess were required to change to dress blues for each evening meal. Noncommissioned officers were permitted to invite wives or girlfriends as guests for Sunday dinner.

Entertainment was available at the American and liberty Theaters and five Knights of Columbus Halls or in the library or bowling alley at the Camp Johnston site. The people of Jacksonville opened their homes to the young men at Camp Johnston. They came in unselected lots and few betrayed the trust. "Invite a soldier to dinner" was a standard slogan. Overseas, the soldiers sent back to Jacksonville greetings and souvenirs. Postcards and letters came from over-there, often with regularity. But sometimes communication suddenly ceased - and the reason later made apparent in published lists.

By June 1918, an additional $1.7 million expansion program was submitted for the camp and would have built quarters and facilities to train 50,000 men.

The project was canceled when the war ended November 11, 1918, and within two weeks, the camp was ordered demobilized. By February 1919, only a few soldiers were left to guard the site.

During WWI, the commanders at Camp Johnston were: Col. Fred L. Munson until April 8, 1918; Col. Charles L. Willard, April 8 to September 29, 1918; and Maj. Gen. William P. Duvall, September 29, 1918, until the close of the war.

In 1919 and 1920, Jacksonville became a motion picture capital. The Chamber of Commerce, in an effort to attract more studios to Jacksonville, established a motion picture committee in 1920. Headed by former "Metropolis" editor W. R. Carter, the committee on January 1, 1921, announced that "the world's largest motion picture production center," would be built at Camp Johnston at Black Point. It would be called the Fine Arts Center and would have offices, film laboratories, warehouses, and costume houses.

Plans were also in the works to construct sets consisting of an "old western town, ghetto streets, a Chinatown and a country village." The funding for the project, provided by Murray W. Garsson of Fine Arts Pictures, Inc., collapsed. Louis J. Selznick then decided to take on the project and he signed Klutho Studios to design the facility. As construction began at Camp Johnston, the picture, "When a Man Deceives," was produced there. Soon after this picture was released, the entire project folded.

On January 4, 1921, representatives of the Loyal Order of the Moose inspected the site for possible selection for a Moose Center. A site on the river at Orange Park was eventually selected, and "Moosehaven" is still located there today.

By 1921, most of the surplus personal property was given to the state of Florida and stored in huge warehouses at Yukon. A fire of unknown origin burned one warehouse to the ground on June 23, 1921. The fire started under an outside platform, and fire fighters thought they had extinguished it at 2:40 in the afternoon. But at 3:20, it was discovered that the south end of the building was ablaze.

The Jacksonville Fire Department was called to the scene only to discover there were no water connections available in the area. They attempted to fight the fire with chemicals, but to no avail. The entire structure was declared a total loss. Also lost were 300 vehicles and parts stored in the warehouse. The vehicles and parts had already been given to the State of Florida, and the state and federal government shared the $1.5 million loss. Firefighters were successful in saving two other nearby warehouses.

Three days after the fire, the federal government gave the state title to 682 acres and 154 buildings. The remaining 458 buildings and supplies were sold at public auction for a fraction of their value. Until 1924, other than annual state militia training encampment, little else happened on the property. For instance, on August 4, 1923, National Guard troops from across Florida assembled at Camp Johnston south of the city for two weeks of instruction ordered by Adjutant Gen. J. Clifford R. Foster. At that time, Col. Raymond C. Turck of Jacksonville commanded the camp.

By 1925, plans were again being made to turn part of what is today NAS Jacksonville into a subdivision for a growing Jacksonville, part into a state campground, and part into a state park.

The housing subdivision, Camp Johnston Park, was platted on October 11, 1925. It covered the approximate area covering today by Mustin Road to the east, Albermarle Avenue to the north, Birmingham Avenue to the south, and Biscayne Street to the west. Already platted, on February 1, 1923, was Casa Linda Colony. This small subdivision encompassed 21 lots, all being located around Casa Linda Lake (then called Lake Elizabeth). The lots were sold and houses and a church built. Both of the above subdivisions were replots of an earlier 1905 subdivision called Peninsula Park.

POSTCARDS OF CAMP JOHNSTON

PARADE, CAMP JOHNSTON, FLA.

GROUP, HUT, CAMP JOHNSTON, FLA.

MOTOR TRANSPORT AND Y. M. C. A. HUT NO. 3, CAMP JOHNSTON, FLA.

BANDSTAND, CAMP JOHNSTON, FLA.

JESS WILLARD AND JACK PHELAN, CAMP JOHNSTON, FLA.

Interior of Barracks at Camp Joseph E. Johnston.

Typical Night Crowd, Army Y. M. C. A. at Camp Joseph E. Johnston.

"Getting into Form" at Camp Joseph E. Johnston.

"Foxing" at Camp Joseph E. Johnston.

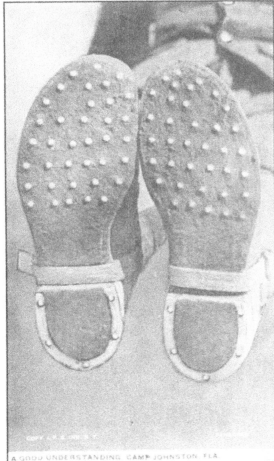
A GOOD UNDERSTANDING, CAMP JOHNSTON, FLA.

ST. JOHN'S RIVER IN FRONT OF CAMP, THREE MILES WIDE HERE, CAMP JOHNSTON, FLA.

OFFICERS' SCHOOL, CAMP JOHNSTON, FLA.

INTERIOR OF CAMP LIBRARY, CAMP JOHNSTON, FLA.

CAMP JOHNSTON LIBRARY, CAMP JOHNSTON, FLA.

20

On June 18, 1928, Major General J. Clifford R. Foster, the Adjutant General of Florida, whose actions in 1905 led to the establishment of military activities at Black Point, died while in office. Shortly after his death, the campgrounds at Black Point were officially named Camp J. Clifford R. Foster in his honor.

Construction to improve the site continued every year through 1938. Typical yearly improvements can be seen in the Reports of the Adjutant General of the State of Florida. In 1929, a swimming pool was built, the west boundary of the camp was fenced, 100 tent floors were constructed, eighteen temporary wooden buildings were taken down and the material was salvaged for other buildings, roads were repaired and numerous other small projects ensued. Also, under the Act of Congress of June 7, 1924, war trophies (cannon captured during WWI) were

placed at the main entrance and other sites at Camp Foster.

One event remembered by long-time Jacksonville residents was the Sunday afternoon car race held on the campgrounds. The races usually attracted sizeable crowds as the cars raced around the camp roads. Also, the first air show was held in the mid-1930s on the campgrounds. A highlight of the show was a Ford Tri-motor doing a complete loop in front of the crowd.

In the early 1930s, the depression set in at Jacksonville. So desperate were the times for the unemployed, the Duval County Citizens Relief Committee in December 1932 turned Camp Foster into "an unemployment, relief and concentration camp." Mayor John T. Alsop stated, "Jobless men who have been begging on the streets will be given an opportunity to enter the camp, or provided with a way to return to their homes. If they do not want either, they

will be sent to the city prison farm." In 1935, in order to beautify the camp, 90,000 pine trees, 450 palms, 100 camphor trees, 50 oaks and 800 oleander plants were planted between Mustin Road and the St. John's River. Except for a few oak trees, virtually none remain today.

In 1935, Jacksonville also started fighting for the Army air base which was to be located in the southeast. Camp Foster was seen as a prime site. Tampa also thought it had an excellent site. After years of hard fought political activity, Tampa proved to have the more experienced politicians and the Army Air Base, known now as MacDill Air Force Base, was located in Tampa.

Jacksonville was not out of the picture yet though. The Navy also started working for a new site for an air-training base in the Southeast United States. Numerous cities campaigned for the air base, and again Jacksonville was in a tough political fight.

Camp Foster front gate. (Florida Times-Union)

Camp Foster Post Cards. Above, lowering the flag. Below, Main Gate.

The story of how Jacksonville finally secured a Navy base is told very well in an article by Lowell Clucas published in the Jacksonville Journal on October 15, 1940, the day the air station was commissioned. Here is that story. Corrections and new information have been added that were classified in 1940 and unavailable to the writer.

In 1934, as Adolf Hitler and Benito Mussolini were starting their military buildups, a group of Jacksonville men took up a discussion of national defense. At the time, Florida had a naval air station at Pensacola, but no army air defense bases.

Led by Major Raymond W. Cushman, the group introduced a resolution to the Chamber of Commerce in February 1935. The resolution deplored the inadequacy of the national defense, stressed the vulnerability of Florida's 1,200 miles of coastline, and urged the President and Congress to provide a system of air

defense with strategically located bases. While it referred to possible bases in Florida, the resolution did not specifically mention Jacksonville as a prospective site, although Major Cushman discussed the matter at the same time before the chamber's board of governors.

As far as is known, this was the first occasion at which an air defense base for the city was brought up for open discussion by a public organization. Soon afterward, Major Cushman left Jacksonville for Washington to attend a meeting of the house military affairs committee on a bill directing the war department to determine bases for air defense. He submitted a brief of Jacksonville to the department, but Jacksonville was eventually excluded from the half dozen localities decided upon.

During most of 1935, however, the air base was looked upon as an army, not a navy project. In December 1935, Carl Vinson, chairman of the House Naval Affairs Committee, announced his intention to come to Jacksonville to view an undisclosed "naval project." Vinson delayed his Florida trip, but a month later it was learned that Miami had offered the government 31,000 acres of land for an air defense base. Meanwhile, Congressman Joe Sears was

busy on the draft of a bill providing for the acquisition of land adjacent to Camp Foster for a proposed Naval Air Station. A hearing on the bill was held April 14, 1936, by the House Naval Affairs committee.

Congressman Sears and Major Cushman were the chief spokesmen for the bill, which gave the government the Camp Foster area without cost. Hardly had Sears finished his arguments, however, when a report from Acting Secretary of the Navy W. H. Stanley arrived, throwing cold water literally on

the whole idea. "The Navy Department has no use for this site either now or in the near future, and as it considered that the expenditure of naval funds in the development of the site would not be warranted, certainly for many years to come. The Navy Department recommends against the enactment of the bill," he wrote.

Nevertheless, the committee reported favorably on the bill a week later. For two years the air base issue hung in mid-air. In the meantime, Tampa acquired the army air base. The

75TH CONGRESS
3D SESSION

H. R. 10031

IN THE HOUSE OF REPRESENTATIVES

MARCH 25, 1938

Mr. GREEN introduced the following bill; which was referred to the Committee on Naval Affairs and ordered to be printed

A BILL

To authorize the acquisition of lands in the vicinity of Jacksonville, Florida, as a site for a naval air station and to authorize the construction and installation of a naval air station thereon.

1 Be it enacted by the Senate and House of Representa-
2 tives of the United States of America in Congress assembled,
3 That the Secretary of the Navy be, and he is hereby, author-
4 ized to accept on behalf of the United States, free from
5 encumbrances and without cost to the United States, the
6 title in fee simple to such lands as he may deem necessary
7 or desirable on the Saint Johns River in the vicinity of
8 Jacksonville, Florida, approximately fourteen hundred acres,
9 as a site for a naval air station to be returned to the grantor if

Sears bill did not become law and while Jacksonville plugged tirelessly away at the project, the matter was finally put in the hands of a special naval board under Rear Admiral A. J. Hepburn.

THE HEPBURN REPORT

Created by Congress in May 1938, the six member Heapburn Board was charged with the study of the Navy's needs. Its purpose, as stated in Secretary of the Navy letter of June 7, 1938, was to "investigate and report upon the need, for purposes of National Defense, for the establishment of additional submarine, destroyer, mine, and Naval air bases on the coasts of the United States." The board was officially called the "Statutory Board on Submarine, Destroyer, Mine, and Naval Air Bases, 1938."

As the board convened in mid-1938, three major air bases were to be established. The following cities and areas were visited for the southeast air base site: Miami, Fort Lauderdale, Key West, Jacksonville, Banana River, Fernandina, Brunswick, Savannah, Parris Island and Charleston.

The board announced its plans to scrutinize Jacksonville and arrived there September 12, 1938. It was met, wined and dined by a large local committee and visited the municipal airport, Eastport, Fleming Island, Green Cove Springs and Camp Foster. Of the five sites, the board was obviously struck by Camp Foster, particularly by the width of the river off Black and Piney points. It was also noted that when the matter of aircraft carriers came up, Rear Admiral E. J. Margaret, one of the board members, remarked that a turning basin might be dredged near the St. John's River jetties. From then on Jacksonville was in the front lines of Florida cities mentioned in connection with a great Southeastern Naval Air Base.

One of the first indications of the favorable wind blowing Jacksonville's way was the arrival October 10 of three seaplane patrol bombers from Coco Solo in the Canal Zone. Under the Command of Lieutenant Commander W. S. Tomlinson, the patrol boats took off and landed for five days off Camp Foster. Some concern was expressed by the aviators over the hyacinths in the river, but apparently the Hepburn board was told that Jacksonville was more than suitable for patrol bomber operations. In any event, Admiral Hepburn wrote

the chamber that the flyers were particularly impressed by the courtesy shown them and then announced his plan to revisit Jacksonville.

This second trip of the board to the city was shrouded in secrecy. The officials arrived November 8, 1938 and so mum was the word that the story did not leak out until January 5, two months later. Even then it came out in a special report of the chamber's aviation committee. At this second conference two additional briefs were submitted to the board. The supplementary reports dealt with a proposed carrier basin at the mouth of the St. John's River and the land in the Camp Foster area.

By November 23, the board assured Jacksonville authorities the information necessary was complete, but that it could say nothing regarding the final selection of a site. On Saturday, October 1, 1938, Mayor Blume left for Washington to call upon members of the Florida delegation for support for Jacksonville. It was the first time since being elected 15 months earlier the mayor had left Jacksonville for more than a weekend.

On December 1, 1938, the Hepburn Board report was submitted. The report stated "The Board recommends the establishment of a major air base at Jacksonville, having the following general characteristics:

(a) Facilities for two carrier groups (planned with a view to expand to four carrier groups).

(b) Facilities for three patrol plane squadrons (planned with a view to expansion to accommodate six squadrons).

(c) Facilities for two utility squadrons.)

(d) Facilities for complete plane and engine overhaul.

(e) Berthing for carriers at inner end of entrance to jetty.

(f) A channel to permit tender berthing at piers at Camp Foster.

(g) Development of an outlying patrol plane operating area in the lower Banana River."

On January 3, 1939, the good news finally arrived in Jacksonville. The Hepburn board recommended establishing an air base at Camp Foster and a carrier basin in Mayport. By January 13 the Jacksonville real estate board agreed to appraise the land free of charge. Steps were being taken to obtain the land.

Legislation was pending before Congress for funding by January 20, and topographical maps were being

prepared on February 2, 1939 using Ford Tri-Motors from NAS Pensacola.

However more grief and hard work was yet to come. Apparently asleep at the switch for months, Miami suddenly announced that its facilities would be far better than Jacksonville's, and that it had plenty of statements to prove it. One member of the board, Captain A. L. Bristol, admitted in a cross examination by Representative Pat Cannon of Miami that Miami possessed strategic advantages superior to those in Jacksonville.

It soon became apparent that a $15,000,000 item in the $65,000,000 Vinson Navy bill providing for the Jacksonville air base would be struck by the House Naval Affairs Committee. The bill was sent to the Senate with the Jacksonville base eliminated.

Clamors for the base were also being directed at the committee by Brunswick, Georgia, and Fernandina, Fort Lauderdale and Cocoa in Florida. On February 12, the House Naval Affairs Committee flew into Jacksonville looked at the Camp Foster site, and flew out again to Miami. But neither city was promised a thing.

Finally Chairman Vinson asked the Navy for more information, and Assistant Secretary of the Navy Edison reconvened the Hepburn Board. The secretary of the Navy directed by SECNAV Precept, QB (119)/P16-3 (380520)G, dated February 23, 1939, stated that the sites again would be evaluated and a supplemental report completed. The earliest possible final report was requested.

The board reconvened in Washington on February 27, 1939, and plans were made to again visit the sites. This time each site was evaluated as to advantages and disadvantages using the following criteria: (a) location, (b) seaplane facilities, (c) landplane facilities, (d) berthing for ships, (e) supplementary operating areas, (f) weather (g) magazine area, (h) water, (i) transportation and, (j) any other characteristics.

In the meantime, the state armory board agreed to give up title to Camp Foster to Duval County providing it should be properly compensated for doing so. This served to untangle the involved property situation at Camp Foster and spiked the principal gun being fired at Jacksonville by Miami. It was also beginning to appear that $15,000,000 would go a lot further in Jacksonville than in Miami where dredging and filling operations at Key

Biscayne were required. The Miami air base site would have increased the cost several million dollars.

With Jacksonville strengthening and organizing its forces with a new Air Base Council under Brigadier General Sumpter L. Lowry, Jr., the Hepburn board prepared once more to visit this city. After a visit to Miami, the board announced it would arrive March 2. Jacksonville in the meantime was carrying on its fight with Miami in enemy territory. A series of advertisements appeared in Miami newspapers condemning the establishment of a naval air base there. Directed primarily at the women of Miami, the ads warned that such a base would wreck real estate values in the resort section. They were signed "For the Protection of Homes."

Jacksonville prepared far in advance for the third tour by the Hepburn Board and the house committee. Streams of publicity accompanied their arrival. The wives of the congressmen were feted in St. Augustine and Ponte Vedra. A banquet at the Roosevelt Hotel in honor of Admiral Hepburn was attended by 300 guests. "No city has ever surpassed Jacksonville in hospitality," Chairman Vinson remarked.

The day after the banquet the visitors sailed up the St. John's River in pleasure boats headed for a luncheon at Camp Foster. There, armed with maps and charts, Commander B. M. Fortson and Major Cushman presented arguments for Camp Foster as the logical site for the base.

The visitors left Jacksonville and reassembled in Washington on March 7. Deliberations continued, and finally on March 22 the announcement was made. The Hepburn Board had reaffirmed its previous recommendation. Jacksonville was the preferred site for the Southeastern Naval Air Base.

Dated March 21, the final decision stated, "The Board unanimously and strongly recommends that this base be established at Jacksonville, Florida." The main advantages of the Jacksonville site included superior strategic location, immediate availability of the site, comparative freedom from hurricane damage, superior transportation facilities, and local labor and industrial resources.

One of the principle items against Miami were petitions from Coconut Grove residents against the air base. As the Board met, 187 protest petitions were received. The primary objections to the Key Biscayne site included noise, restriction of yachting and fishing areas and depreciation of property values.

Since the Army/Navy first visited Jacksonville in 1917 to establish an air base, the dream was becoming a reality. From this point on no further opposition arose against constructing the air base at Jacksonville.

Senator Andrews on March 27 amended the naval air bill passed by the House to the Senate to provide for a $17,000,000 major base at Jacksonville, an auxiliary station at Banana River and carrier berthings at Mayport. Following suit, the House Naval Affairs Committee approved the supplementary report of the Hepburn Board on March 28 and concurred in the Senate amendment.

An elated Jacksonville delegation including Gilchrist Stockton, Mayor Blume, General Lowry, Commander Fortson, Major Cushman, City Commissioner Thomas Imeson, County Commissioner James G. Cary and Ted Bayley of the junior chamber was in Washington to hear the news. How modestly they regarded the base at the time may be observed in the prediction that 500 or 600 men would be employed during the construction program, that Navy personnel would total 2,000, and the payroll was placed at $5,000,000 a year.

The $66,800,000 omnibus authorization bill for air base appropriations was passed without debate by the Senate on April 19, and the House approved it the next day and sent it to the White House. President Roosevelt signed the bill April 25.

Approved now for keeps, the Naval Air Base went to the engineers of the Navy Department for construction plans. Again the program was remarkably modest in comparison to what the future held. Three years were planned for construction of the $15,710,000 station. $1,000,000 was to be spent at Jacksonville during the first year, 1939-1940. This was to be stepped up to $4,000,000 during 1940-41 and to $10,171,000 during 1941-42. 1939-1940 saw more than $21,000,000 in contracts; $6,000,000 more than the combined total of the original three-year program. Since the government had been assured from the start that the county would turn over lands adjacent to Camp Foster at no cost, action was speeded up to acquire the properties. County Attorney J. Henry Blount was directed to write a bill creating the Duval County Air Base Authority. The authority was to be composed of the county commissioners and with the concurrence of the citizens in the community, empowered to issue bonds and to levy taxes to pay for them. The bond money was to be used to pay for the land acquisition. The state legislature approved the bill, and the newly established Air Base Authority held its first meeting on May 13, electing Robert D. Gordon as its chairman.

The first meeting should have been conducted with an aura of optimism surrounding it. The day before the citizens of Duval County expressed their attitude towards the new base with a jubilee celebration. Highlighted by parades and a flyover by Navy aircraft, the celebration included such miscellany as boxing and wrestling bouts, roller skating and jitterbug contests. Even a song was written for the occasion.

The money question was first on the agenda for the Authority. July 18 was set as the date for a $1,100,000 bond election. 100 prominent local men and women were appointed to publicize the details of the election, and speaking programs were organized. Quiz sessions were held on the radio, and the registration books were opened. Commander C. H. Cotter, the engineer commissioned to build the station, arrived in Jacksonville and urged that speed was essential in acquiring the Camp Foster land since not a shovel could be turned until the acquisition was assured.

As the election neared, dozens of announcements appeared in the press containing endorsements of the bond issue by various civic clubs. Large advertisements were also displayed. On the appointed day, in spite of a pouring rain, Jacksonville citizens surged in and out of the polling booths, and by nightfall the election had been won hands down. The final vote was 13,808 for and only 265 against. It was one of the most one-sided bond issues ever voted for in Jacksonville history.

92.1 percent of the registered voters cast their ballots. At least fifty per cent of the registered voters (7,639) "had" to vote and also vote in favor of the bond issue to secure the base. The problem of Camp Foster itself remained to be solved. No decision had been made as to its monetary value or when it should be relocated. A bitter argument ensued between the leaders of Jacksonville and the armory board, not over compensation, but over the camp's future site. Jacksonville wanted to keep the camp in Duval County, but the armory board favored a Clay County site between Keystone Heights and Starke. The armory board prevailed and received $400,000 for Camp Foster.

On October 8 bonds were placed on sale and fifteen days later construction began with the dismantling of several Camp Foster buildings for movement to Clay County. Throughout the winter thousands of dollars in contracts were awarded, and April 1941 was discussed as a possible commissioning date for the new base.

By the end of June 1940 it was evident the base would be ready for commissioning before the April 1941 date. One thousand workers were busy on the site, and Commander Cotter announced that contracts awarded between October and June totaled $8,000,000; eight times the amount planned for the first year's construction. A detachment of Marines had arrived from Parris Island, South Carolina, to police the growing project, and Captain Charles P. Mason was announced as the future first commander of the new air station.

Then on July 12, it was announced by the navy department that a giant $12,786,000 cost plus fee contract had been awarded two Jacksonville firms and a Georgia concern for the erection of dozens of additional buildings. It was by these three, George S. Autchen Company and Duval Engineering and Contracting Company of Jacksonville and Ratson-Cook of West Point, Georgia, that the building of a trade school and cadet training area were constructed. Shortly afterward, the Navy department also reported it would establish an auxiliary naval station at Green Cove Springs. Naval Air Station at Jacksonville was now well on its way.

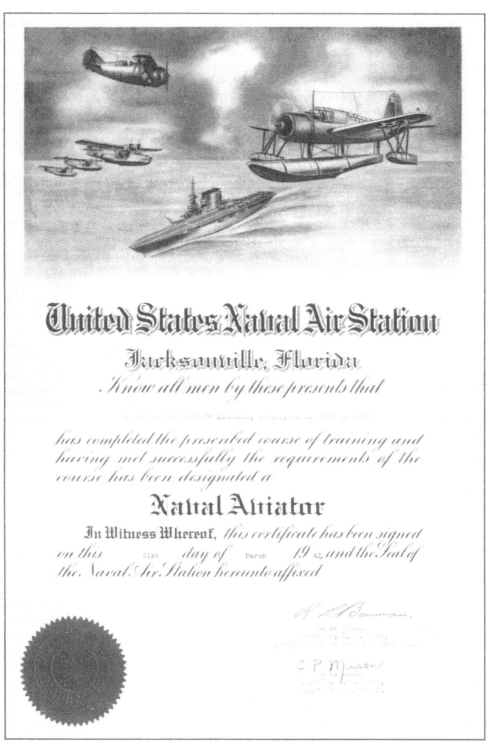

LAND ACQUISITION

As 1940 approached, construction of the new Naval Air Station had started. Land acquisition, thought to be a relatively easy process, turned out, in actuality, to be much more difficult. A development to be located where Kent Campus of Florida Community College of Jacksonville now sits, was awaiting a decision of a taxing litigation case which was before the Florida State Supreme Court.

The State of Florida lost this case and the 300 proposed housing units could not be placed on the property tax rolls—much to the chagrin of taxing officials. Other property settlements concerning the main station were not faring much better. Camp Johnston Park and Casa Linda Colony subdivisions were both heavily populated, and some of the residents were not pleased at the thought of condemnation of their land.

Condemnation suits commenced, however, beginning on November 23, 1939. The first suit included the old Camp Johnston (Foster) site and one lot of the Philbrofen subdivision. Next, condemnation started in the Sherwood Forest subdivision (now the officers' housing area). Additional condemnation suits followed to encompass the entire present day base and the Dewey Park area.

The naval air station, which today lies in Township 3 South, Range 26 East, came into existence through condemnation suits which were heard in the United States District Court as follows:

a) Suit 123-S-Civil - Encompassed 1939.8 acres and $388,744.73 was paid to the owners.

b) Suit 124-J-Civil - Encompassed 10.87 acres and was settled for $4,250.

c) Suit 157-J-Civil - Encompassed 99.5 acres and $65,186.24 was paid to the former owners.

d) Suit 164-J-Civil - Encompassed 98.84 acres and was settled for $46,452.90.

e) Suit 167-J-Civil - Encompassed 105.97 acres and was settled for $38,336.

With the State of Florida deeding all state highways located within the compound to the United States Government, the main base land was starting to be secured. Another suit acquired 112 acres for $18,000 for the Dewey Park Housing Project. As of July 11, 1940, however, 41 of the 243 parcels condemned were still to be settled. As of September 9, with an October 15 commissioning date approaching, the only part of the air base site still not in litigation was the original Camp Foster 1,000 acre site. In October, the nearly 1,000 acre site owned by Mr. Zaring was threatening to go to appeals, and it looked as if the land payment haggling was going to go on for years. But eventually all sides agreed to a price and all of the land was purchased for $934,250. An additional $1,583 was also spent in court costs. The land at Ribault Bay (Mayport) was tackled next while construction came into full swing at NAS Jacksonville.

Title to the entire 3,260 acre reservation was vested back on April 23, 1940. This was in spite of the fact payment to the owners was not finally settled until late October.

THE COMMISSIONING

Captain Charles P. Mason arrived in Jacksonville on September 17, 1940, and set up his residence at the George Washington Hotel. A 49 year old native of Harrisburg, Pennsylvania, he was designated naval aviator number 52 after receiving training at Naval Air Station, Pensacola, in 1916 and had enjoyed a variety of assignments during his naval career.

He organized the Navy's first seaplane patrol squadron in the Pacific Fleet and had been executive officer of the Navy's first aircraft carrier, USS Langley, and of the USS Yorktown. He also supervised the construction of the first USS Saratoga.

Ashore he served as executive officer of Naval Air Station, Norfolk, and was commanding officer of Naval Air Station, Key West. He was experienced in the training of young aviators having served as a flight instructor and twice as Superintendent of Naval Aviation at Pensacola.

After serving 19 months at the Naval Air Station, he would go on to command the aircraft carrier USS Hornet and retire as a rear admiral. He was promoted to vice admiral on the retired rolls and settled in Pensacola

The station was commissioned October 15, 1940 at noon. Shown in the picture are RADM John Towers, CAPT Charles P. Mason and CDR C.H. Cotter. Building 110, the first completed structure, is shown in the background. (Florida Times-Union)

where he served several terms as that city's mayor.

Because of the superb planning ability of Commander Cotter, the base was ready for commissioning much earlier than initially projected. Captain Mason recommended October 15, 1940 as the ceremony date, and by October 4 his recommendation was approved by Secretary of the Navy Frank Knox and Rear Admiral William H. Allen, Commandant of the Sixth, Seventh and Eighth Naval Districts. With war raging throughout Europe, Congress had passed a National Defense Program and Naval Air Station, Jacksonville, would be the first base to be commissioned under that program.

It was a Tuesday morning when Captain Mason assembled his 21 officers and 42 enlisted men for the ceremony. The sparse Navy contingent was joined by an Army bomber squadron from MacDill Field in Tampa, the Marine corps band from Parris Island, South Carolina, Jacksonville's city fathers, Senators Claude Pepper and

Charles O. Andrews and hundreds of Jacksonville citizens. Among the proud organizers of the movement to acquire the air station were Gilchrist Stockton, Ted Bayley, General Lowry and Commander Fortson.

At 11:59 a.m. a bugler sounded "Attention", and Commander Cotter stepped to the podium. He opened the ceremony with a description of the construction program declaring, "The station has construction projects underway worth more than $20,000,000."

After praising the cooperation he had received from the Navy departmental, city and county officials, and the contractors, he introduced Vice Admiral John Towers, Chief of the Bureau of Aeronautics.

Admiral Towers expressed appreciation for "the remarkable achievement Commander Cotter has performed in getting the station ready for commissioning in such a comparatively short space of time." He then congratulated Captain Mason on his new command describing him as

one of the very few officers specifically qualified to take the assignment.

Commander Cotter then stepped forward and said, "It gives me great pleasure, Captain Mason, to turn this station over to you."

Captain Mason expressed his appreciation to Commander Cotter, the contractors and workmen, read his orders from the Navy department and declared, "I hereby designate the Naval Air Station Jacksonville, Jacksonville, Florida, in commission." He then turned to his executive officer and ordered, "Hoist the colors." When the national flag had been raised, he ordered, "Set the watch."

The entire ceremony was over in 20 minutes to ensure minimum description of construction activities. The official party and designated guests moved from the field in front of Building 110 to inside the building for a luncheon.

Construction continued at a rapid pace, and by December 1 all base departments were in operation getting ready to start full training activities on January 2, 1941.

Supply Building 110 construction, early 1940. (U.S. Navy)

CONSTRUCTION ACTIVITIES

When the state militia completed its move from the old Camp Foster site to the new Camp Blanding area, only three structures were left standing. One of these remains today, that being a section of the former Navy Exchange complex at Enterprise and Langley Streets. So complete was the move that most of the manhole covers were taken. The majority of the buildings were torn down. Some were reassembled at the new site, and some were moved to Camp Blanding entirely intact. It was now time for the site to start growing as a navy base.

The Jacksonville station was originally planned as an operating and repair base, but developments in Europe in 1939 changed the mission to include training. Construction got under way on October 25, 1939 with the awarding of the first contract covering dredging and filling of the area now occupied by Hangars 116 and 117 and the northeast portion of the landing field, Operations Building 118, Hangars 122, 123 and 124, the Assembly and Repair Building 101 and Maintenance Building 136.

The first facility to have construction started was Pier 139, as test pilings were driven in January 1940 and a connecting railroad was built. The first building to have construction begin was the Supply warehouse Building 110 (now the Fleet Industrial Supply Center) followed by Seaplane Hangars 123 and 124, the permanent barracks and the first nine officers' quarters.

By February 1 ten contracts had been awarded totaling $1,853,000. Bids were accepted to build 16 huge gas tanks; however bids were being rejected for the officers' quarters. The low bid of $150,000 was $30,000 over the funding allocated for the project. The land now surrounded by Yorktown Avenue, Allegheny Road, Mustin Road and approximately Birmingham Avenue was an almost impenetrable jungle until the end of 1940 when the efforts of approximately 3,500 W.P.A. (Works Progress Administration) workers finally cleared the area.

The gatehouse was designed in February and by March 13 the impressive architectural drawings for the $1,000,000 Airplane Overhaul Shop were shown. Of the 700 buildings constructed during WWII the Assembly and Repair Shop was the most massive and costliest. To expedite construction on the mammoth 1.5 million square feet

Construction of Cadet Training Facilities, 1940. (Florida Times-Union)

of floor space facility, the foundation, steel frame and superstructure were contracted to three separate companies. Construction was started in July 1940. It remains today the largest art deco structure in Duval County.

The seaplane hangars (now home to the helicopter squadrons under Helicopter Squadron Wings Atlantic) and the old torpedo workshop were designed by one of America's great architects, Albert Kahn. Construction on Hangars 122 and 123 started in late March and on the barracks and mess halls in April. Also in April, bids were being accepted for the land plane hangars (114, 115, 116), station maintenance building (103), fire station and garage (105) and laundry facility.

As construction inside the base was proceeding, so were plans for a new road to the base. Numerous landowners donated right-of-way for the establishment of what is now Roosevelt Boulevard. Washington deemed construction of a bridge at Ortega for the road necessary for national defense.

Labor relations gave base officials problems in May, bringing some projects to a virtual standstill. Steel workers walked off the job, protesting the 56-hour week they were required to work. The steel workers, receiving $1.25 an hour, wanted a 40-hour work week. The problem was solved when Commander Cotter required the contractor to start working two separate shifts.

Construction of the assembly and repair shops, the radio control tower, old parachute loft, garage and fire station,

maintenance building 1376, roads and walks, Hangar 117, paint and oil storehouse, Boiler Plant No. 1, Water Plant No. 1, Boathouse, and the gasoline distribution system began in May 1940. The operations building and control tower were started in June and the senior bachelor officers' quarters and the administration building were started in July.

On June 28, 1940, a cost-plus-fixed-fee contract was awarded to three companies for general construction. This contract, N04-4132, was active until 24 April 1943, during which time the three contractors were paid $48,394,266 for construction of the Naval Hospital, Naval Air Technical Training Center's initial 30 buildings, the auxiliary air stations at Cecil Field, Mayport and Lee Field at Green Cove Springs and other outlying fields.

From May 1943 to September 1944, sixty-three more lump sum contracts were issued. As of April 1942, $68,123,842.10 was spent on construction activities. As the construction progressed, it was not without problems, however. Most of the underground steam distribution system had to be replaced within one year due to electrolysis. Absenteeism on the construction sites continually slowed production, along with wage rates that varied from contractor to contractor for the same craft.

On the landing field, grading operations were started about the middle of August 1940 and 75 train carloads of limerock a day were pouring

Construction of Hangar 124. (U.S. Navy)

onto the field in September. The field was entirely completed early in December. Constructed at a total cost of approximately $1,000,000, the field in 1940 was one of the largest in the country with runways averaging 5,800 feet in length and 300 feet in width. The longest runway was 6,000 feet.

October saw the completion of the Supply Warehouse, Building 110. Shortly thereafter, on October 14, 1940, nine houses in the officers housing area were completed. Supply operations commenced on October 16 when Commander T. E. Hipp, the Supply Officer, moved into his offices. Construction of the 200-bed base hospital commenced in mid-November.

As work was finishing in Hangar 114, the station recorded its first fatality when a painter fell 26 feet to his death. Accidents continued to plague the station and in 1943 a safety engineer was hired from California to help solve the high accident rate problem.

In July 1941 the housing complex at Park Street and Roosevelt was completed. During September, steel for hangars now used by the Navy Exchange and Auto Hobby shop started to pour in. The hangars original use were utility hangars for the Technical Training Center. The Officers Club formally opened September 30, 1942 and less than one month later, on October 25, 1942, a $30,000 fire closed it for at least two months. St. Edwards Catholic Chapel was dedicated on

January 18, 1943 and in March the All Saints Chapel was dedicated. Although construction continued until the end of World War II, the majority of the station's 700 buildings were completed by the summer of 1943. But two years earlier, the station was active accomplishing its primary purpose of training personnel for the impending war.

STREET NAMES

Initial roads in the area the station now occupies were laid out in 1906-1907 when the Peninsular Park and Philbrofen housing developments were plotted. Those in Peninsular Park were named after presidents, and a variety of names were used in Philbrofen. Among them were Priscilla, St. John's, Selma and Francis. Although none of the original street names are used today, the basic road pattern does.

The first marked roadway names were Camp Johnston Road (now Albemarle), Old Jacksonville Road (officially designated State Road 3 and now Allegheny), and Riverside (now Mustin Road).

With few exceptions streets of the new air station were named after ships. Magnolia Drive inside the new weapons compound was named in honor of Magnolia Plantation, a pre-Civil War plantation located above the banks of the St. John's River south of Black Point.

Mustin Road was named in honor of pioneer Naval aviator Henry B. Mustin.

Allegheny Road was named for the Navy tender of the same name that was assigned in Pensacola in 1940. Enterprise Avenue, previously designated State Road 169 and planned as the main roadway into the station, was named for one of the Navy's first line aircraft carriers.

Streets that penetrated the new industrial area of the base from the west were named with one exception for aircraft carriers or tenders either in service or soon to join the fleet. The exception was Birmingham Avenue named after the heavy cruiser.

Albemarle, Barnegatt and Curtis were named after seaplane tenders. Lexington, Saratoga and Yorktown took their names from aircraft carriers. Ranger Street was named in honor of the first Navy ship commissioned by John Paul Jones.

Many of the north-south oriented streets were similarly named with Hornet, Langley, Wasp and Wright named for aircraft carriers, and Ballard and Ajax named for new destroyer tenders. The tender class had just recently been redesignated from fleet mine sweepers. This class of ship was named after birds, hence the station names of Avocet, Gannet, Heron, Lapwing, Owl, Sandpiper, Swan, Teal and Thrush. In 1974, all roads established in the then new housing area, south of Naval Hospital Jacksonville, were named using the letters of the alphabet, starting with "A." Later that year, they were renamed after current station operating squadrons. These new roads included the names Hurricane Drive (squadron VW-4); Pelican Drive (VP-45); Shamrock Drive (HS-7); Batman Drive (VP-24); Pro's Next Drive (VP-30); Red Lion Drive (HS-15); Eagle Run (VP-16); Trident Court (HS-3); Sea Horse Court (HS-1); Dragon Killer Court (VP-56); and Fleet Angel Court (HC-2). The last street unofficially changed was Patoka in front of building one. Capt Delaney changed it to ASW Lane (Antisubmarine Warfare Lane) in keeping with the stations mission of antisubmarine warfare. A new street was added into the hangar built for VP-30 in August 1996, but is yet to be named.

The first aircraft assigned to NAS Jacksonville was a Grumman J2F "Duck" Amphibian flown to the municipal airport from San Diego on January 17, 1940. This plane was also the first to land at the base. On March 22, 1940 with Commander V. F. Grant at the controls, it landed at the seaplane ramp with no advance fanfare, picked up Commander Cotter and flew out to check construction activities at Banana River. On April 2 the plane was flown to Naval Air Station Pensacola to have a two-way radio installed.

Eventually two more aircraft, including Captain Mason's private plane, were also housed at the Municipal Airport. The first landing on the NAS Jacksonville runway site was on September 7, 1940 at 3:04 P.M. A Naval Aircraft Factory N3N-3 landed with Commander Grant again at the controls and Commander Cotter as his passenger. The landing was accomplished just three weeks after clearing operations for the new runway commenced. After a brief 20-minute stop, the engine was started, and Commander Grant flew back to Municipal Airport. In reality, this was the third recorded landing at the air base, the second being a landing one evening by Commander Grant on the St. Johns River near the seawall during a severe thunderstorm some two weeks earlier.

Sunday, October 13, 1940 marked the permanent assignment of four aircraft to the base proper. The three aircraft from the Municipal Airport arrived with a newly assigned fourth, much to the chagrin of Captain Mason. The commanding officer restricted the aircraft operation from the base on October 27, because, "It seems that every time a Navy plane takes off from the station, the whole working populace has a tendency to lean on the shovel, so to speak, and watch the war bird fly away." But by the end of December, operations became common occurrences with 30 aircraft in the station's inventory. The last nine of the new N2S-3s arrived December 30.

Thursday, January 2, 1941, 25 new N2S Stearmans took off to begin flight-training operations. P2Y-3 seaplanes started appearing in February, followed by SNJs in March and April. The PBY-1 and 2 Catalinas started appearing in June, replacing the P2Y's. By July 1941 ten air station pilots were enroute to San Diego to fly back Ryan NR-1 training planes. This was a tedious task, as a route back to NAS Jacksonville had to be planned to allow for a gas stop every

First assigned airplane, a J2F Grumman Duck, sits on the ramp at the Jacksonville Municipal Airport, January 1940. (U.S. Navy)

The first aircraft to land on the new runway was this N3N-3 on September 7, 1940. The pilot was CDR Grant, Station XO and his passenger was CDR Cotter, the first Public Works Officer. (Florida Times-Union)

200 miles. By December 1941 a total of 578 aircraft were assigned to the station as follows:

Grumman J2F 1
Beech JRB-2 1
Grumman JRF-5 1
Naval Aircraft Facility N3N-3 33
Ryan NR-1 99
Stearman N2S-1 114
Stearman N2S-3 94
Vought OS2U-2 53
Vought OS2U-3 42
Consolidated PBY-1 21
Consolidated PBY-2 10
Curtiss SNC-1 19
North American Aviation SNJ-2 5
North American Aviation SNJ-3 85

This inventory remained fairly intact until the last class of aviation cadets graduated on February 26, 1943. With the graduation of these last ten cadets, only air operational advanced training was conducted, and the NR-1 Ryans and N2S Stearmans aircraft were discontinued from service at NAS Jacksonville. As the trainers departed other aircraft started arriving to take their place.

The TBF Avenger and OS2U Kingfisher appeared during 1942 and the F4U Corsair and F6F Hellcat in 1943. By 1943, the Assembly and Repair Shops (now NADEP) were working on the SNJ, OS2U, SNB, F4F, FMI, F2A, TBF, F4U, TBD, SBD, PBY, GBI and SB2U models of aircraft. During 1942 and 1943 two landings and two takeoffs were occurring every minute at the airfield.

The first Superintendent of Aviation Training, Commander R. L. Bowman, successfully set up and developed the aviation-training program during his 19 months of duty. As head of the Aviation Training Department, he spent from September 30, 1940 to January 1941 organizing the department. As the first training flight took place the first primary training squadron was established. On February 3, 1941, intermediate training using the PBY Catalina was started, and the first cadets completing intermediate training were commissioned as Naval Aviators during August 1941.

As intermediate training was introduced, the SNJ "Texan" aircraft was also used, and the main station field had to be enlarged. On July 1, 1941, the area known as the MAT was completed which provided an additional 2000 feet of paved area in any direction. A 300-foot circle was painted in the center of the airfield enabling traffic to take off to the left of the circle while other aircraft landed to the right of the circle. At the same time planes made touch and go landings through the center of the circle. All the early training was of the landplane type but on March 5, 1941 two seaplane squadrons using P2Y's were established. Commander J. B. Dunn assumed the duties of Superintendent when Commander Bowman became the Executive Officer of the station, May 25, 1942. Commander Dunn's tour of duty was of short duration, lasting for only three weeks. After the Japanese attack on Pearl Harbor, guns were obtained

and installed on the stations PBY-1s and PBY-2s (Catalinas) and antisubmarine patrols commenced on April 1, 1942 along the east coast with these aircraft. Also in April 1942 the Naval Air Operational Training Command was established at Naval Air Station, Jacksonville. Although operational training flights would not commence at NAS Jacksonville until August, the NAOTC controlled the operational flight training at eleven other air stations. In June the Staff Utility Pilot was added to the activities of the Aviation Training Department. The station Operations Officer, Commander R. Goldthwaite, USN, was chosen to be the third Superintendent of Aviation Training. It was during his tour of duty, which began June 13, 1942, that plans were made to change the training from Primary and Intermediate to Operational. Primary Training was completely discontinued in August 1942, and the entire training program became Intermediate. Plans for Operational Training progressed and on October 15, 1942, VPB2 OTU (Patrol-Bomber Operational Training Unit) #1 became the first Operational Training Unit. Throughout the remainder of 1942 the Intermediate program was gradually discontinued and new Operational Squadrons formed. Additional VPB2 squadrons were formed as follows: VPB2 OTU #2 on April 5, 1943, VPB2 OTU #3 on June 3, 1943 and VPB2 OTU #4 on September 1, 1944.

Two VB (Bomber) Operational Squadrons, which were formed at NAS Jacksonville, were moved to Sanford and Lake City when those stations were commissioned. The Operational Training Program was rapidly expanded and additional squadrons were formed during the term of Commander R. R. Johnson as Superintendent from November 22, 1942 to June 28, 1943. The first of these new squadrons, designated T-4, was established to train landing signal officers, and the second was an air bombing training unit known as T-5. These were followed by a VO-VCS (Observation, Cruiser-Scouting) Squadron flying OS2U "Kingfisher" aircraft and a VTB (Torpedo-Bomber) squadron flying the TBF "Avengers." Early in 1943 the training of cadets was discontinued and the station was henceforth devoted to Operational Training.

On April 28, 1943, a Tower Duty Officer was assigned the responsibility for the control of all land plane traffic.

Since the student load had become so much heavier, it was decided training in some of the larger squadrons would be more efficient if there were two squadrons conducting the same type training. Two new squadrons were established as VPB2 OTU #2 on April 5 and VPB OTU #3 on June 3, 1943.

Commander R. R. Johnson was appointed executive officer of the station on June 28, 1943, and was succeeded as superintendent by Commander B. E. Moore, who had served as squadron commander of VPB2 OTU #1.

Several changes were made in the training while Commander Moore served as superintendent. The two VTB OTUs moved to Miami to take advantage of more extensive gunnery areas, and the air bomber unit moved to Banana River. In August 1943, Lieutenant Commander T. B. Snook was assigned as Assistant to the Superintendent for Statistics and Personnel.

The executive officer, Commander Johnson, served as Acting Superintendent after Commander Moore was detached on December 8, 1943 and until Commander John Raby reported onboard on January 10, 1944.

On October 26, 1943 VF (Fighter) type of operational training began to replace the TBF (Torpedo-Bomber Fighter) training which ceased at NAS Jacksonville on November 12, 1943 and moved to another operational training command base. VF-5 was established

on November 18, 1943, and training of pilots in fighter tactics using the F4U Corsair was commenced. With the beginning of operations of the speedy Corsair, course rules had to be altered slightly, and additional types of strafing targets and seaward gunnery areas had to be adjusted since gunnery was one of the main functions in fighter training.

In March 1944 the first Beech SNB aircraft was assigned to the station, and multi-engine instruction was inaugurated. In July 1944 VO-VCS Training was discontinued when OTU #1 moved to Pensacola.

Naval Air Station Jacksonville established a remarkable record for the first time since the training program began, completing the entire month of June without a fatal crash. The record was remarkable considering that there were 26,000 hours flown in operational training that month, and the majority of the hours were flown in the Corsairs. Exclusive of seaplane operations, the month of June showed an average of 1100 landings and take-offs a day—the highest day being 1585. This made a general average for the month of one landing and one take-off every 35 seconds during the entire period of daylight flying. It must be considered there were slack periods and peak periods, and some landings and take-offs were conducted simultaneously. It also should be considered that during a normal day there were a number of emergencies, which slowed the operation of aircraft, accounting for

slack periods and the resultant build-up.

During this same month, transient aircraft on flight plans averaged 103 per day—the highest day being 146 inbound and outbound flights making a monthly total of 3,203.

After an extensive study, the two VPB OTUs were consolidated in August 1944 to obtain greater efficiency, reduce duplication of training, decrease the personnel assigned and facilitate all phases of training. On September 1, 1944, VF OTU #4 became MF OTU, the first commissioned Marine squadron at NAS Jacksonville, and also the first in the Naval Air Operational Training Command.

VPB-OTU #4 was commissioned at the same time.

A new type of training was added to the Department about this time when the Communication Training Unit was formed to train instructors in communications for the various operational training units. On September 10, 1944 an outstanding milestone was attained, when the station aircraft passed one million hours in flying time. Over a period of 45 months, a total of 9,324 students had been trained since the first student began training in 1941.

On December 1, 1944 the OS2U Kingfisher station ready plane was replaced by an SNJ-2. On December 15, an SNB-2H Ambulance Plane was received for use by Naval Hospital Jacksonville and by December 30 the OS2U station weather plane was

The first training flight takes place on January 2, 1941. (Florida Times-Union)

replaced by a SDB-5.

A most important unit of the Aviation Training Department was the Aviation Cadet Regiment. This regiment was formed with the commissioning of the naval air station in October 1940. The Officer in Charge, Lieutenant Commander Roger W. Cutler along with Lieutenant John R. Yoho executive officer, and Lieutenant S. W. Carr as his regimental adjutant, laid plans for the organization of the Cadet Regiment. The first class of cadets arrived in December of 1940, a group of about fifty men.

By February 1941 cadets were arriving at the rate of two classes of about one hundred men each per month. The cadets had approximately three months training in ground school and primary flight training at various Naval Reserve aviation bases and were designated as Seamen Second Class, V-5. They continued in that status until completion of a two-week course of indoctrination in the cadet regiment. The indoctrination included a study of Naval Customs and Traditions, marlin-spike seamanship, and related courses. At the end of this period, they were sworn in as Aviation Cadets and began their ground school course.

For the first several months regimental headquarters was located in one wing of a cadet barracks. The growth of the regiment and the increase in the number of administrative officers made it necessary to move to larger quarters. With a fitting ceremony, including an address by Captain Mason, the cornerstone to the new regimental headquarters was laid in April 1941. The building was completed in July of that year.

Toward the end of April 1941 the first two cadets to complete primary training George Snortlidge and Clifford Hemphill were commissioned as Naval Aviators. A group of twenty-nine in the first class were commissioned on May 5, 1941. By July 1941, the cadet training program was increased significantly. A Navy program designed to train 8,700 cadets using Pensacola, Jacksonville and Corpus Christi, Texas was increased to 10,200. The last class of aviation cadets at NAS Jacksonville was awarded their wings on February 26, 1943. The largest class ever graduated at one time was 118. Some 4,363 aviation cadets came to Jacksonville as fledgling airmen and left as trained, efficient Navy pilots with a complete knowledge of aeronautics, after flying some 760,010 hours of training. Great feats were performed by these same men in air action against the enemy in the Pacific. Some even returned to Jacksonville with decorations for heroism and courage in battle.

In March 1945 the station airfield had a lighting system installed, and PBY-5As were added to Squadron VPB OTU #1.

Aviation activities started to decline rapidly after the war was over. Squadron VF-5 was dissolved July 30, 1945 and satellite training fields were closed in October 1945. The Air Operational Training Command transferred out and the NAATC followed in February 1948. The base was starting to resemble a ghost town. Training in PBYs ended November 13, 1945 and by November 1, 1946, the great PBY fleet was gone. But VO-VCS seaplane training commenced again on June 10, 1946 with the Seahawk. As the training was starting, a freak hail storm hit the auxiliary station at Cecil Field, damaging 46 aircraft — three beyond repair. To keep its newly trained pilots, the Naval Air Reserve was established at Cecil Field on June 30, 1946. In September, headquarters was moved to NAS Jacksonville, where it remains today. The Naval Air Advanced Training Command had three assigned squadrons at Jacksonville in 1947. These were Instructors ATU #9, VA, ATU #4, and VO, ATU #6. The last PBM Martin Mariner, which had flown at NAS Jacksonville for years, finally departed on April 22, 1948. On May 27, 1948, sixteen student pilots received their wings as Naval Aviators. This was the first time since 1943 a class had been conducted for student training. On June 17, two Phantom Fighters appeared as the first jets to land at NAS Jacksonville drawing huge station crowds. On November 1, Fleet Air Jacksonville was established and Cecil Field was reestablished. Carrier Air Group 8 was also established at NAS Jacksonville. Commander Air Force Atlantic Training Unit #1 was establishing at NAS Jacksonville on December 1, 1948. The purpose of the unit was to further train former carrier pilots. Two additional carrier groups were assigned to Fleet Air Jacksonville in February 1949. Carrier Air Group 4, consisting of the only jet squadron on the Atlantic coast in VF171, which was assigned to NAS Jacksonville and Carrier Air Group 17, which was assigned to Cecil Field. Carrier Air Group 8, in the meantime, was in Washington participating in the largest aircraft parade in history at President Truman's Inauguration. By April 7, 1949, NAS Jacksonville was considered "The Navy's war plane capitol of the country." Some 500 combat planes and 10,000 men were assigned to air groups at NAS Jacksonville and the auxiliary station at Cecil Field. A Philippine Mars arrived on April 21, 1949, the largest seaplane ever to land at the station. By the end of 1949, Carrier Air Groups 8 and 13 were to be lost but VP3, VP5 and Fleet Air Wing II were reassigned to make up the loss. The VP Squadrons, flying the P2V-3 Neptune, would arrive in 1950. NAS Jacksonville had fought the air war with its well-trained pilots. It had also gone from wood and canvas to becoming the Navy's East Coast jet center in nine short years. The forties had ended and little did air station personnel know aviation would continue with the Korean Conflict in the near future. But through the efforts of numerous citizens in Jacksonville, Fleet Air was commissioned along with Patrol Squadrons, which still exist today. The base's aviation future was secure.

This N3N-3 (on floats) was used by student pilots for primary seaplane training. (US Navy)

FLIGHT TRAINING

Training activities started on January 2, 1941, with the first instruction flight. Flight training was initially only of the primary and intermediate types with advanced flight training added on May 6, 1942. Primary training was given to pilots, most of whom were required to have at least two years of college. The entire flight training course lasted about seven months, and 200 pilots a month were to be turned out.

The first two weeks were spent in "indoctrination" in the cadet regiment. Indoctrination consisted of becoming familiar with Navy routine and customs. Following this were two weeks of ground school learning instrumentation of navigation, airplane engines, radio, etc.

After a student completed his primary and basic landplane courses, he would then go to one of three specialized training groups. One group specialized in catapult planes, with final assignment to a seaplane detachment onboard a battleship or cruiser; the second group was in patrol planes, and the third group in carrier-based landplanes. Except for a few who were retained as additional flight instructors, students joined the fleet after completing training.

Initial flight training was conducted in the Stearman Trainer (known affectionately as the "Yellow Peril" to students because of the numerous accidents associated with it) and in July 1941, the Ryan NR-1 was added. The PBY Catalina was used in intermediate training.

Commander Bowman, brought in from NAS Pensacola to establish the flight training program, also established an instrument training program using the Link Trainer. The Link Trainer consisted of an airplane fuselage equipped with short wings and tail surfaces that responded to control, but anchored to the ground. This trainer pales in comparison to the electronically sophisticated trainers located on station today, but it served its purpose well. A hood was usually pulled over the cockpit of the trainer, and the student was required to bank and turn using instruments. Later he was taken aloft

with an instructor and taught to do the same maneuvers in an airplane.

Operational flight training commenced May 6, 1942. The primary purpose of the "post graduate" sky-fighting school was to teach newly-winged pilots and air crewmen the final maneuvers that prepared them to go directly into combat against the enemy. From the beginning numerous outlying air fields were necessary for flight training. Fields at Ft. Lauderdale, Daytona Beach,

Melbourne, Deland, Sanford, Vero Beach, Lake City, St. Simmons Island, Georgia, and auxiliary air stations Lee Field and Cecil Field and auxiliary air field at Mayport all were part of the Air Operational Training Command.

Three aviation gunnery schools were also in operation, at Jacksonville, one in Hollywood and one in Purcell, Oklahoma. The arrival of the gunnery school also saw the arrival of the famous F4U Corsair. The Corsair along with the F4F Wildcat practiced simulated carrier landings at Switzerland Field, near Green Cove Springs. The F4F was being soundly defeated in dog fights with the Japanese Zero. The development of the F6F started in June 1941 as a "pinch hitter" for the F4U. The F4U had just been accepted by the Navy, but numerous problems had to be ironed out prior to combat flying. The F6F Wildcat was developed by Grumman to replace its older brother — the F4F. The F6F was developed and accepted by the Navy almost without change by July 1942. This aircraft appeared at NAS Jacksonville shortly thereafter. This aircraft, with its anti-bullet protection, ended WWII with a win/beaten ratio of 49:1 against the Zero. This exceptional aircraft was also used by the Blue Angels as their first aircraft when they formed at NAS Jacksonville in 1946.

Meanwhile at NAS Jacksonville, torpedo bombers and dive bombers were making similar training flights at the auxiliary landing fields. Switzerland Field was used for limited bombing practice but its use eventually was stopped when a pilot almost bombed the east end of the Shands Bridge at Green Cove Springs by mistake. While pilots practiced carrier landing tactics, landing signal officers were taught simultaneously on the ground. By

October 1, 1944, some 6,338 operational training students had been received with 4,798 graduating. Operational flight training was not without its dangerous side, however, as 58 fatalities were recorded in the first 29 months of operation.

The flight training conducted at NAS Jacksonville during WWII could be divided into the following areas:

Primary Training in NSNs, NR-1s and N2S2;

Intermediate Training in SNJs, SNCs and OS2Us;

Patrol Bomber Intermediate and Operational Training primarily

in PBYs;

Torpedo Bomber Operational Training in TBFs;

Scout-Observation Operational Training in OS2Us.

Landing Signal Officers were trained using F4U-1s and F4Fs.

Although flight training was the principle training activity,

other training for the demands of war were also conducted.

SYNTHETIC DEVICES

All of the 6,338 operational training students who received instruction at NAS Jacksonville had in their curriculum about 10 hours instruction in synthetic devices. Their time in link trainers and celestial link trainers included study of fighter direction, homing, celestial navigation and straight navigation. In addition fighter pilots received about 10 hours instruction in a device known as the Gunairstructor, which supplemented the actual gunnery phase of their flight training.

All aircrewmen who underwent operational training were given instruction using a device known as the 3A2 Free-Gunnery Trainer, which served the same purpose in training aircrewmen as the Gunairstructor did for the fighter pilots. Actually thousands of combat aircrewmen, graduates of free gunnery schools, were given supplementary training at NAS Jacksonville using such synthetic devices as the 3A2 Gunnery Trainer. Each man received approximately 15 hours of instruction. Combat aircrewmen were also given training in power turrets and in mock-up devices such as tunnel guns and waist guns.

ENLISTED TRAINING

In addition to pilot and aircrew training, the station provided training for thousands of enlisted men who kept the planes flying. Every other month examinations were given for advancement in aviation ratings. Strikers and candidates for advancement trained in the squadrons or in the Assembly and Repair Department, and those who made sufficient progress were recommended for advancement. Each second month, five to six hundred such advancements were made.

Most of this training was for aviation machinists mates, aviation radiomen, aviation ordnancemen and aviation metalsmiths, but significant training was accomplished for such ratings as aviation electricians mates, aviation radio technician, aerographers mates, photographers mates, aviation storekeepers and parachute riggers. Thousands of men who held general service and specialist rating but who never worked in aviation activities, had received their training and advancement at NAS Jacksonville. Before the formation of the Naval Air Technical Training Center (NATTCEN) as a separate command, NAS Jacksonville conducted class "A" service schools, training strikers, for AD, AMM, ARN and AON ratings. Several thousand men graduated from these schools before the NATTCEN was established as a separate command.

The Supply Department also acted as a school for strikers for the commissary, storekeeper and stewart branch ratings; the Medical Department was a training school for hospital apprentices striking for pharmacists mates; the post office was a school for mailmen strikers; and thousands of storekeepers and yeomen had received training in the various offices of the station. The Yard Department trained seamen and coxswains for duty on rescue and salvage boats along with firefighters for the Specialist (F) rating; in fact it was the job of every department and every division on station to train enlisted personnel for ratings at the same time they carried on their daily work.

GROUND SCHOOL

An extensive ground school course of the type at Pensacola and Corpus Christi was conducted at NAS Jacksonville throughout the period when primary cadet training was carried on. The course was designed to supplement the flight training program in producing an all-around naval aviator. The syllabus included power plants, structures, communications, gunnery, navigation and aerology.

INDOCTRINATION SCHOOL

Approximately 300 A-V(S) officers received their indoctrinational training at NAS Jacksonville in a school established by the Chief of Naval Operations in January 1943. These officers played a major part in carrying out executive duties throughout the command. Indoctrination of officers was also accomplished in the Assembly and Repair Department. Training was provided in Assembly and Repair organization and procedures, and graduates were ordered to various assembly and repair departments throughout the Navy.

INSTRUCTORS' SCHOOL

The Instructors' School was conceived in early 1944 by Commander J. S. Thach, training officer for Naval Air Operational Training Command, and Lieutenant Commander D. B. Thornburn to standardize pilot instruction materials and method of presentation. Standardization of instruction materials stemmed from the requirement of the Fleet for pilots uniformly trained in tactics and procedures to afford more closely coordinated combat teams in Fleet units. The Instructors' School aided in meeting this by providing written lectures on various phases of operational training. The schools also served as a clearing house for new developments in combat flying which were evaluated and passed on to the operational training units.

Standardization of presentation was achieved by having new flight instructors attend the school's eight-day course. They were taught the latest methods of instruction embracing psychology, salesmanship and the use of graphic visual aids. Classes were conducted in a conference setting to encourage discussion, criticism and suggestions.

NAVAL ACADEMY GRADUATES

The 1944 and 1945 graduating classes of the Naval Academy at Annapolis received a short post-graduate course in naval aviation at NAS Jacksonville to make them familiar with Navy planes and aeronautical equipment.

RECRUIT TRAINING

A recruit training station was operated at NAS Jacksonville from September 10, 1942 until August 1, 1944. Approximately 20,000 recruits were screened, given primary indoctrination, issued clothing and received inoculations. They were also classified for Class "A" Service Schools and special programs and received rudimentary training in seamanship.

VOCATIONAL TRAINING

A major training program was also established on the station for civilian personnel. An extensive in-service vocational training program was established in September 1941 for the training of civilians in the aircraft trades. By the end of December 1944 more than 4,500 civilian employees, including many women, had received training under the guidance of the Vocational Training Division in the Assembly and Repair Department. Students from Robert E. Lee and Andrew Jackson High Schools were also recruited for vocational training.

Supervisory or instructor training courses was given to 1,225 mechanics. In addition, training programs were established in the Supply, Transportation, Inspection and Survey, and Public Works Departments. During the month of December 1944, approximately 1,200 civilian employees were receiving on-the-job training in the various civil service trades.

LOW PRESSURE CHAMBER

One of the more interesting phases of the training carried on at NAS Jacksonville was accomplished in the main dispensary. Pilots and aircrewmen were trained for high altitude flying using devices that changed air temperature and pressure and simulated conditions at various altitudes. Approximately 60 men per day were given oxygen and high altitude indoctrination providing training as well as a complete laboratory for experimentation with the effects of high altitude flying on aviators. Although this work was in an experimental stage, many improvements in aviators' comfort and safety resulted.

AVIATION SUPPLY OFFICERS' SCHOOL

Another valuable course in the aviation training program was carried on at NAS Jacksonville at the Aviation Supply Officers' School. This school was established by the Chief of Naval Air Operations Training to meet the fleet need for supply officers especially trained to procure, store and issue aeronautical material. The school was conducted by the Supply Department with more than 525 supply officers graduating.

Approximately one half of the time was devoted to observing disassembly and repair of aircraft engines at the Assembly and Repair Department. The curriculum included courses in aviation indoctrination, aviation supply, aeronautical parts and material, instruction in the basic principles of aircraft, aircraft recognition and in aviation gunnery.

PHYSICAL TRAINING

The Physical Training Department was established on July 1, 1942. Physical and military training programs were developed for all naval personnel aboard the station. Officers and men in flight training received primary consideration of the department, but programs were carried on simultaneously for officers and men permanently assigned and those in other training programs. These programs were developed with three objectives:
(1) Survival Skills
(2) Physical Conditioning
(3) Morale maintenance through organized recreational activities.

Training in survival skills was conducted in a jungle-like area near where the Marina is today. Survival training was a feature of the program since its beginning with special emphasis given on swimming skills, the use of regulation aircraft emergency equipment and general emergency procedures.

COOKS AND BAKERS SCHOOL

At the request of the Training Division, Bureau of Naval Personnel, a Cooks and Bakers School was established at NAS Jacksonville. Instructors were recruited from the civilian community, and the school had 240 students in four classes of 60 each. Students received both classroom work and practical experience in the station galley.

COMMISSARY STEWARDS SCHOOL

In the fall of 1942, a Chief Commissary Stewards School was organized at the request of the Training Division, Bureau of Naval Personnel. The school was established to prepare ship's cooks and bakers, first and second class, for advancement to the rating of chief commissary steward. The school carried an enrollment of 120 men and the instructors were all chief commissary stewards with many years experience.

NAVAL AIR OPERATIONAL TRAINING COMMAND PROJECTS

In addition to the training units listed above, there were a number of other important training units operated by the Naval Air Operational Training Command. These activities included the aviation gunnery training standardization unit, the Naval Air Operational Training Command training film unit, and the special devices shop, administered by NAS under direction of the Instructors' School.

Physical training is conducted at the station in this August 1942 photo. Note the PBY in the St. John's River in the background. (U.S. Navy)

JACKSONVILLE NAVAL AIR STATION
CHERRY POINT MARINES

PRICE
10c

JACKSONVILLE NAVAL AIR STATION
vs
CORPUS CHRISTI "COMETS"

PRICE
25c

JACKSONVILLE NAVAL AIR STATION
vs
MIAMI N. T. C.
MUNICIPAL STADIUM
JACKSONVILLE, FLORIDA

PRICE
10c

JACKSONVILLE NAVAL AIR STATION
FT. PIERCE "AMPHIBS"

PRICE
25c

Although NAS Jacksonville was mainly concerned with the business of war in the early 1940s, the private and social life of personnel on the station was not forgotten. Most of the officers and student officers lived along Mustin Road just south of Yorktown. Near the corner of Yorktown and Mustin Road was the Student Officers area and their social life centered around the very large swimming pool between the buildings. Just south of the student officers quarters was the Wave's area where about 1,200 officer and enlisted Waves lived.

Just west of this area were the Chapels where services are still held today. Across from the chapels on Mustin Road were the five Junior Officers Quarters and further south on Mustin Road was the Officers Club where some of the social life for the officers of the station still takes place

today. But in the 40s, there was music and dancing on Wednesdays and Saturdays.

South of the Officers Club two-hundred Senior Officers lived in the Senior Bachelor Officers Quarters (BOQ) which was generally regarded as one of the most beautiful and attractive Senior BOQs in the whole Navy. Officers of the rank of Senior Lieutenant and above were eligible for Senior BOQ as long as there was space for them.

South of the Senior BOQ was the Naval Hospital and the quarters for the Senior Medical Officers. Still further south, between the hospital area and the circle, there was about a mile of river front homes, quarters for the Senior Staff Officers of the Naval Air Operational Training Command, and Senior Officers of the Naval Air Station and the Naval Air Technical Training Center. While most of the station was given over to the

business of war, Mustin Road was maintained as a beautiful thoroughfare. Most of the really attractive buildings were located between Mustin and the St. Johns River and it was this area which gave NAS Jacksonville the reputation of being one of the most beautiful of all the Naval Reservations.

During this period any officer, commissioned or warrant, was eligible to use the Officers Club, as long as he was not on student officer status. Since all training at NAS Jacksonville by 1943 was operational in nature, many of the officers who were in a student status had the rank of lieutenant or captain (marine). Regardless of rank, they were not eligible at the Officers' Club while undergoing operational training. Instead, they were given use of an adequate student officers club, a two-story frame building facing Yorktown, which was maintained in Building 700.

The 1942 NAS Jacksonville "Fliers" football team. (U.S. Navy)

Dances were held at the Student Officers Club on Saturday nights and hard liquor was served with the limitation that no officer on flight duty the following day was allowed to drink.

A large two-story brick building stood at Allegheny Road near its intersection with Albemarle. This had been a country store and filling station before the station was built when what is now Allegheny Road, was Old Jacksonville Road. This building was renovated and attractively decorated as a Chief Petty Officers Club. It soon became the center of social life for the "Chiefs" living in the Station Housing development. Popular entertainers were brought to NAS Jacksonville, as were movies.

The second station newspaper still called today the "JAX Air News," issued its first edition on April 1, 1943 at the recommendation of then Commanding Officer, Captain Michael. The name stuck and today this weekly paper is one of the most widely read in Jacksonville. It replaced the original station newspaper, the "JAX FLYER" which stopped printing in March 1943.

By May 1943, fifty-six softball teams had been organized and were playing on the station's seventeen diamonds. Nine pools were opened and billiards, bowling, basketball, volleyball, badminton, a golf driving range, and a hand-ball court were available.

By October 1943, Joan Leslie had won the station's "Pin-Up" girl contest, soundly beating out such favorites as Betty Grable, Ann Sheridan, Gene Tierney and Lana Turner.

In November 1943, first run movies were shown at the station theater, with Mickey Rooney's "Girl Crazy" heading a long list of hits that brought movie attendance to 1,000,000 yearly. At first, the movies cost a nickel, but some months later, even this small admission fee was abolished.

Championship fights known, as "Smokers" were popular events while the base's basketball, football, and baseball teams (all named the "Fliers" attracted enthusiastic participants and regularly defeated opponents.

The football team, which started in 1941, was coached by Lieutenant Commander Edward W. Mahan, one of Harvard's all-time football greats. Their outstanding team won 7 out of 8 games and played its last game on December 6, 1941. The following day Pearl Harbor was attacked effectively terminating the 1941 football season.

The 1942 football team began a truly memorable season with a win over the University of Florida "Gators," and ended it with a win over Duke. This team was ranked 6th nationally with a 9 and 3 season.

NAS Jacksonville was not permitted to have a football team in 1943 due to local regulations, but the NATTC Air Raiders represented Jacksonville and won three games of a seven game schedule. The team was formed again in 1944, and football lasted throughout the rest of the 1940s.

The basketball team was rated number 11 in the nation in 1945 with the University of Kentucky at number 4.

In addition to sports, station personnel were also concerned about the war effort, leading all others in the nation for the sale of War Bonds in January 1944.

The station attracted its share of celebrities during this period. Actress Gail Patrick met and fell in love with Lt. Arnold White, a naval flier. They were married at the All Saints Chapel in July 1944. In one of the biggest entertainment events of the war, Bob Hope gave the first of his two station performances to date on March 14, 1944. So popular was the "JAX Air News" edition commemorating the event that a second printing was made for the only time in the newspaper's history. Frank Sinatra had also made a little-publicized singing appearance. Other stars that appeared during the war years included Brenda Jolle, William Gargen, Louis Armstrong, Brian Donlevy and Cary Grant.

When the station was commissioned, some of the departments were already in operation and by December 1, 1940, all of them had some organization. Initially, all activities on the station reported directly to the Commanding Officer. The first tenant to break away was the Naval Hospital when it was established in 1941. Other activities followed suit in 1942. Today, there are 74 tenant activities aboard NAS Jacksonville. The following departments were established initially: Aviation Training Department (which included an Operations Division), Ordnance Department, Communication Department, Yard Department, Personnel Division, Supply Department, Medical Department, Public Works Department, Naval Air Technical Training Center and Assembly and Repair Department. In addition to these, there was a Ship's Service Officer, Conservation Officer, Legal Officer and public Relations Officer. Histories of bigger departments are detailed as follows.

NAVAL AIR TECHNICAL TRAINING CENTER

The Naval Air Technical Training Center, the largest activity ever to exist at NAS Jacksonville, has a history that started even before the station was commissioned. On August 10, 1940, Commander Junius L. Cotton, who had received word that he was to be Officer-In-Charge of the proposed schools, wrote a letter to Captain Mason, who was soon to be the Commanding Officer of NAS Jacksonville, discussing in detail the plans for setting up the training activity at the station, to be known as the "Trade Schools". The letter also requested 290 officers and men for office forces, instructors, master at arms, etc. Two months later, on October 15, 1940, the Naval Air Station at Jacksonville was commissioned. "One of the largest activities at the new air station is the trade schools," said an article in a special edition of the Florida Times-Union of that date. By that time plans for the new schools were well advanced, although it was not until March 2, 1941 that classes actually convened in the Aviation Metalsmith and Aviation Machinist's

Mate Schools. On April 1, the Aviation Radioman and Aviation Ordnanceman Class A Schools were added.

"The trade school buildings cover about six city blocks and at present consist of thirty buildings," erected as temporary structures to meet the emergency. These buildings consisted of "thirteen barracks to house as many as 3,000 men; one administration building, six classroom buildings, one bakery, one mess hall, one auditorium, six shop buildings and one land plane hangar." An auditorium, known now as King Hall, accommodated 800-900 students and a swimming pool was provided for recreational purposes. Six one story structures housed equipment such as lathes, drills, grinders, heat treatment ovens, welding tables, plating equipment, etc. for instructional purposes. Graduates were to be assigned to no shipboard duty except in connection with the naval aviation service. "In this respect," concludes the article "the trade school at Jacksonville is unique for there is today no other school in the country where so many specialties will be taught and where the men are being trained exclusively for aviation duty."

The trade schools were organized as a unit of the Naval Air Station with Commander Cotten as Officer-in-charge and with the Commanding Officer of the Naval Air Station as his immediate supervisor. Indicative of the humble beginnings is the fact that the Officer-In-Charge of the Aviation Radio School, Lieutenant Commander MW-Wells, also acted as Executive Officer.

No major changes occurred in the administrative program until September 22, 1941, when Lieutenant Commander (later Commander) Ronald D. Higgins, relieved Commander Cotten as Officer-in-Charge of the Trade Schools upon Commander Cotten's detachment to Quonset Point, Naval Air Station, as Executive Officer. In October 1941, the Trade Schools became a separate command under the Chief of Naval Aviation Training, Chicago.

The schools had been growing. The December 1941 roster showed a total of 3,160 students and 53 officers, It was in February 1942, under the drive developed since Pearl Harbor, that real development and growth in activities

began. That month showed 21 new barracks under construction to be completed at three per month, with corresponding messing and school facilities. The Marine Aviation Detachment was created in February with Major William T. Evans as Commanding Officer. Also in February 1942, in order to avoid confusion with the many trade schools that were springing up all over the nation in answer to war needs, and to point the way to the more technical and scientific aspects of the training, the name "Trade Schools" was changed to "Service Schools." The addition of facilities, doubling the capacity of the schools, reflected the constant demand for trained men.

On January 1, 1943, the Chief of Naval Air Technical Training at Memphis assumed full jurisdiction of the command and on February 10, 1943, the name "Service School" was changed to "Naval Air Technical Training Center" and the center became a tenant activity of NAS Jacksonville. The school had been operating on a two-shift, 6-day a week schedule since March 1942, and continued this schedule until July 1943.

By the time the war ended, the following schools had been established: Aviation Metalsmith, Aviation Machinist's Mate, Aviation Radioman, Aviation Ordnance-man Class A, Aviation Ordnance Officers', Aviation Fire Control, Aviation Radar Operators, Bombsight, NT School (Aircraft Turrets), Aviation Electrician's Mate, Mobile Ordnance Instruction Detachment Units, Aviation Storekeepers, Searchlight Maintenance, Automatic Pilot, Aviation electrician's Mate and Crash Fire and Rescue Training. The battle cry for each school was "keep 'em flying." The school, throughout the war, served its main purpose of providing technically trained personnel to Naval Aviation in such quantities as were required. On February 13, 1948, the order to "cease instruction" was given and three days later 2,500 personnel and tons of equipment were moving to Memphis for a consolidation of the school. The operational flight trainers remained here along with two of the schools. By May 1948, the Technical Training Center compound resembled a ghost town. The Technical Training Center was eventually to return to NAS Jacksonville in the 1950s.

ASSEMBLY AND REPAIR DEPARTMENT

The Assembly and Repair Department was commissioned in 1940, along with the station itself with Commander Crinkley as the first Assembly and Repair Department Officer. His first task, after obtaining the facilities, was to get trained aircraft workers and equipment—much of which was specialized. In October 1940, the Commandant at Norfolk was directed to train a nucleus of 50 men with the understanding they would be transferred to NAS Jacksonville on January 1, 1941. These men, along with some aircraft mechanics transferred from NAS Pensacola, formed the original work force. The embryonic Assembly and Repair Department was scattered over the station and an administrative office was maintained for a short time in the contractors building. In January 1941, the offices were moved to the second deck of the Administration Building and from here the department was administered until the early part of May. During the months of February and March 1941, the shops with machinery, tools and equipment such as lathes, milling machines, grinders, etc. were set up in the Service School area.

In these early days, the Supply Department had not begun to stock the necessary parts and tools needed for the overhaul of aircraft on the station. So the departments work consisted mostly of getting the tools and equipment necessary to make minor repairs on N2S-1 and 3 airplanes which were used as primary trainers on the station at that time. Mechanics furnished their own tools, "borrowed" what they could from the trade schools, and salvaged parts from airplanes damaged beyond repair to keep planes in the air. Some of the work accomplished in this period consisted of overhaul of SNJ wings, ailerons, and small surfaces; engine buildup; major overhaul of a Continental 400 by Engine Over-haul; and N2S and SNJ propeller overhaul.

In May 1941, the Assembly and Repair Officer, with a clerical force of eight, and the Inspection Group moved into the new Assembly and Repair Department Building. During the months of June, July and August 1941,

the shops were moved, one by one, into the new building. The assembling of machinery, the laying out of floor space and divisional lines, and the classification of personnel was done in July 1941. The first piece of equipment to be set up was a heat-treating furnace. The engine test stands were not completed, so engines were tested on fuselages with the fabric removed. The first major overhaul completed was the Stearman N2S No. 3423 on August 15, 1941, closely followed by a major overhaul of a SNJ. The department consisted of almost 200 employees and a regular schedule of "One a day N2S completed" was introduced at this time.

The year 1942 brought further expansion. Two seaplane hangars were taken over for the purpose of repairing large patrol planes such as PBYs and PBMs which could not be accommodated in the Assembly and Repair Shops, and to do minor repair on fighter type planes. Labor shortages were becoming acute in 1942, so on April 18, 1942, 400 women took the civil service examination for the $ .50 an hour ($26.00 a week) position of mechanic learner. The work shift continued to be 2 shifts, 6 days a week, and this would not end until after the war.

By the end of 1943, the Assembly and Repair Department had received its first F4U Corsair for rework. By 1944, Corsairs were a major part of the workload at the department and there were 4,634 personnel employed. Assembly and Repair Units were also formed in 1944. These consisted of specialized "teams" that went worldwide to correct Naval Aviation problems. As the war ended, the Assembly and Repair Department ended its night shift on November 13, 1945. The normal workday reverted to 0715 to 1615 with an hour for lunch.

In 1947, a new indoor firing range to test overhauled machine guns was opened, replacing the outdoor range, long familiar to Assembly and Repair personnel with its daily machine gun test firing. By February 1947, the R4D Overhaul Program, begun in January 1945, was terminating and full attention was given to the overhaul of 32 Corsairs a month. By March, the Engine Overhaul Division, organized in the spring of 1941 in Building 522 of the "Trade School" area, had expanded. Beginning with overhauls of the Continental 670, used in the N2S Stearman, it had progressed to manufacturing all its own tools and overhauling 31 different types of engines by 1944. By the middle of 1944,

250 R-2800 Pratt and Whitney engines a month were being over-hauled. A record of 287 was completed in May 1944.

The Assembly and Repair Department also set production records for March and April 1947, exceeding the Bureau of Aeronautics aircraft overhaul schedule for the first time. During March, 39 Corsairs, 38 Helldivers and 3 R4Ds were completed. This 80-plane total exceeded by 5 the established production level of 75. In April, 84 aircraft were completed against the goal of 60. This was a far cry from June of 1946, when a goal of 63 was met with a production rate of 10.

In June 1947, a two-week shutdown of Assembly and Repair was ordered. Out of 3,333 employees, 400 remained to help install some new equipment and do a general clean-up. As the Navy was gearing down in June 1947 to a peace-time activity, budget reductions were starting to cause rumors of closure, but the Chief of Naval Operations announced that the Assembly and Repair Department was to remain "a "Class A" aircraft Assembly and Repair establishment." On July 22, 1948, the Bureau of Aeronautics directed use of terminology "Overhaul and Repair (O&R)" vice "Assembly and Repair (A&R)." Also in July 25 Corsairs (FG-10), 3 Hellcats (FGF-5) and 1 Texan (SNJ-6) were repaired.

By March 30, 1948, 450 employees in O&R and Public Works were dropped from the payroll. In order to better relocate shops and functions, an 8X14 foot model of the O&R was also constructed. This model was successfully used for years to correct physical problems. By July 1948, the O&R were hiring additional 800 workers to help alleviate an increasing workload. Minor personnel fluctuations occurred throughout the 1940s, but the O&R Department had certainly contributed significantly to Naval Aviation during WWII.

SUPPLY DEPARTMENT

The Supply Department activities began on September 9, 1940 with the arrival of Commander T. E. Hipp as the prospective Supply and Accounting Officer. Before coming to NAS Jacksonville, Commander Hipp recruited employees for his prospective Supply Department from the Naval Aircraft Factory, the Naval Air Stations at Norfolk and Pensacola, and the Navy Yards at Norfolk, Washington, D.C. and

Charleston, South Carolina. When the station was commissioned, his staff consisted of only fifteen recruited employees and eleven laborers.

On November 1, 1940, the Supply Department moved into Building 110, which had reached a sufficiently complete stage for occupancy. Spare engine parts, aircraft components and complete engines were stored in Building 110, and complete airplanes and salvaged aircraft material was stored in Building 109. An additional Building, 108, was used for the storage of paints and oils.

A great deal of difficulty was experienced in 1941 meeting squadron needs, but this was eventually ironed out. More replacement parts were needed for the N2S Stearman than for any other aircraft. By 1944, the Supply Department was maintaining parts for 24 types of aircraft flying or being repaired at NAS Jacksonville. Buildings 137, 138 and 111 were completed in 1942, and Buildings 160, 162, 1163 and 164 in July 1943.

Under the Supply Department was the Commissary Department. Activities of the Commissary Department increased so dramatically that during the peak of its operations it provided subsistence for approximately 21,000 enlisted personnel.

In 1944, the Station Print Shop reverted from the Communication Department to Supply. By December 1944, the Supply Department was stocking automotive equipment, the original staff of 26 had mushroomed to 1,732, and the department had 869,100 square feet of storage space. The Supply Corps also obtained supplies for 20 other bases. After the war, the department's size was reduced, but the basic Supply mission continued throughout the 1940s.

OPERATIONS DIVISION

The Operations Division was under the Aviation Training Department through out the war years. Under the Operations Division were Aerology, a Utility Unit, a Local Flight Regulation Unit, a Station and Outlying Field Section, a Crash Detail Section and a Navigation Equipment Section. The division supervised all flying, established the flying course rules, directed all rescue activities at crash scenes and guarded all flights departing and arriving the station.

The department also assigned land and sea target areas, scheduled target boats and strafing targets, and laid out field carrier arresting gear. Eventually the Parachute Loft and Photography Department were under Operations Division authority. The first Operations Officer was located in the Naval Air Technical Training Center area. His first tasks were to locate and procure the small outlying fields and to establish course rules and a crash bill. On January 2, 1941, the Operations Division was setup in Hangar 113 at the southwest corner of the airfield and began its primary function, the supervision of flight operations.

On April 5, 1941, the Operations Division moved into its permanent quarters in building 118. The Operations Division was located one deck below the tower to afford a view of both the land and seaplane landing areas. Target boats were procured in 1942 and navigational lights were placed over one hundred miles out in the ocean to teach students how to find a ship or base on a seaward mission. When operational training commenced, the airfield had to come under positive radio control, which meant correct radio procedure had to be followed, traffic around the field had to be controlled and the old system of using visual signals had to be ended. At first, the tower was manned only during actual flight training, but with an increase in transient traffic, 24 hour a day manning commenced late in 1942.

During the time of U-boat activity in the Atlantic, 100 PBY-5As assigned to the station from Fleet Air Wing Five kept in constant communication with the Operations Department. On January 29, 1943 the first group of enlisted Waves arrived and started assuming the duties of control tower personnel. By June 1944, the control tower was responsible for the movement of 603 land planes and 103 seaplanes, controlling all approaches and takeoffs at the airfield and on the St. Johns River. By the end of the war, the department had grown in size, and progressed remarkably in a short period of time to support the aviation activities at NAS Jacksonville.

REMOTE AIRFIELD SITES

In 1940, since pilot training was initially the major function of NAS Jacksonville, remote fields for practicing landings, take-offs and eventually tactics and bombings were necessary. The main airfield at the station would be busy enough just controlling landings and take-offs without any training activities being conducted in the area. So the first task of the newly formed Operations Division was to identify and procure remote areas. The only rule in obtaining the sites was to try to keep them within a 50-mile radius of the station. In addition to the auxiliary stations at Cecil Field, Green Cove Springs, NAS Banana River and Mayport, some 37 other sites were obtained, By January 1946, most of the sites were returned to the local cities and governments, with Herlong being given to the City of Jacksonville on November 19, 1946. NAS Jacksonville has one remote site today, that being Whitehouse, which the station acquired after the closure of NAS Cecil Field. Shown below are most of the remote sites used during the 1940s.

AIRFIELD SITES

Paxson (Duval County)
Trout Creek (Duval County)
Hart Field (Duval County)
Francis
Municipal Airport Tomoka
Switzerland (St. Johns County)
Putnam
Fleming Island (Clay County)
Tomoka
Brannan (Clay County)
Belmore (Clay County)
Herlong (Duval County)
Campville (Clay County)
Keystone Heights Airfield (Bradford County)
Foremost (Duval County)
St. Simmons Island (Georgia)
Fernandina
St. Augustine
Cummer
St. Mary's
Jacksonville Heights (Duval County)
Palatka
Brunswick (Georgia)
Bunnell Naval Air Facility
Pomons (Putnam County)
Bostwick (Putnam County)
Mile Branch (Duval County)
Folkston (Georgia)

BOMBING Sites

Mill Cove (Doctor's Lake)
St. John's River - Great Marsh Island
No Man's Land Island - Nassau Sound
Amelia City (1.6 miles Southwest)
Baker
Black Creek
Bell
Spencer
Chaffee

AUXILIARY AIR STATIONS

Navy Auxiliary Air Station - Jacksonville Municipal Number 1

Commissioned on July 1, 1944, by Rear Admiral Andrew C. McFall, USN, the Naval Auxiliary Air Station, Jacksonville Municipal No. 1, was situated at the Jacksonville Municipal Airport on North Main Street.

This particular station was not activated until the war in the Pacific resulted in significant development in the use of the Liberator bombers. Intrepid Navy fliers demonstrated that the PB4Y-1 was effective in not only carrying out high altitude search missions far into Japanese-held territory but in mast-head attacks on Japanese shipping and low attitude strafing of enemy shore installations.

Taking advantage of these discoveries, the Navy placed more stress on Liberator training and an extra period was added to the training program. Photographic training was also brought within the scope of the instruction.

Municipal Airport No. 1 at Jacksonville offered one of the longest runways on the east coast, plus the advantage of the Florida flying weather. Hence this site was chosen for a training site.

Lee Field Auxiliary Naval Air Station

NAS Jacksonville was quickly overrun with aircraft, so it was necessary to develop additional fields to take the strain off of the station's runways. Benjamin Lee Field - the Naval Auxiliary Air Station at Green Cove Springs, Florida, was formally dedicated as the first addition to NAS Jacksonville on March 12, 1941. It quickly grew in importance as one of the top-ranked fighter-pilot training centers in naval aviation circles.

The base, which included two satellite gunnery fields at St Augustine and Palatka, was located approximately 15 miles south of the parent NAS Jacksonville. The station eventually embraced some 1,300 acres of land and the buildings and public works equipment cost in excess of $8 million.

The station started out as a primary training center for aviation cadets with N2S Stearmans and NR-1 Ryans located there. At one time, Lee Field was the home of one of the first land-based bomber squadrons, and when pre-operational phases were started it was the site of Squadron VN-16, the largest of its type in the nation. In late 1944 Lee Field was used as an operational training base where commissioned student officers were going through their final training prior to going to combat. Pilots were then flying the F4F Wildcat and the F4U Corsair fighters.

On the day the field was commissioned Mrs. Joshua C. Chase, mother of the naval officer for whom the base was named, was an honored guest. Benjamin Lee II was killed in England a month before the end of WWI and was awarded the Navy Cross posthumously. He spent most of his life in Jacksonville.

Commanding Officers during WWII were Commanders Frank T. Corbin; W.H. Ashford, Jr.; Charles Shone and Willard E. Hastings. In late 1945, as the Navy was demobilizing, the first war-weary ships started arriving at the site along the St. John's River, and the station eventually became the home to Atlantic Fleets Reserve Group that grew into 400 ships worth approximately $13 billion by 1954.

Cecil Field Auxiliary Naval Air Station

Cecil Field started out as a copy of the facilities at Lee Field in Green Cove Springs. Aircraft started arriving just before the attack on Pearl Harbor; but operations were stepped-up considerably after the attack. The station was commissioned on April 20, 1943.

Construction actually began April 1, 1941, and was mostly completed by November. The field was originally designated Auxiliary Number 2 of Naval Air Station Jacksonville and consisted of a circular landing mat. It was expanded and runways added in 1943. The station initially consisted of 2,666 acres and was located approximately 16 miles west of NAS Jacksonville. The first landing mat was completed in September 1941 and barracks were ready for occupancy in October. Lieutenant Commander Joseph B. Dun was the first officer-in-charge, from May 21, 1941 until December 1, 1941. Commander Thomas D. Southworth, USNR, succeeded him

and then became commanding officer when the station was commissioned.

Although initially used for primary pilot training, the role of the station was changed to providing the fleet trained pilots and aircrewmen for dive-bombing operations in the Pacific by late 1942. The main planes trained in during WWII was the Dauntless SBD and eventually the SB2C Helldiver.

Mayport Naval Auxiliary Airfield

The site for Mayport was selected at the same time the site for NAS Jacksonville was selected. It was designed to accommodate aircraft carriers to provide pilot carrier training to the pilots at NAS Jacksonville. Dredging of the harbor began in early 1940, but was not completed for over a year. The dredged spoils were used as fill material for the rest of the station and in particular, the airfield. Mayport consisted of approximately 700 acres. Normally approximately 50 officers and 250 personnel were stationed there during WWII. NAS Jacksonville also sent some 50 WAVES to the site for duty. NAAS Mayport was officially commissioned on March 20, 1943.

The station served as a site for aircraft refueling and maintained crash boats and an OS2U Kingfisher, which were used for air-sea rescues of pilots. The first air-sea rescue craft actually arrived in January 1942. T boats were also known to use the facilities for training purposes.

YARD DEPARTMENT

The Yard Department (known today as the Security Department) was first established in November 1940, with Lieutenant C. 0. Bain as the first Yard Officer. The Yard Office was first located in the old Boatswain's Locker, moved in January 1941 to an old contractors building, moved in March 1941 to Building 1, moved in August 1942 to Building 954 and moved in December 1943 to Building 9, at the main gate, where it remained until 1987. At that time the department was split with part staying in building 9 and part going to the second floor of the Weapons building, 876.

The Yard Officer had additional duties as Athletic and Welfare Officer, but with the growth of the station and with more duties assigned to the Yard Department, it was necessary, in August of 1941, to appoint Lieutenant Commander F. W. Mahan, as Athletic and Welfare Officer. Commander F.M. Gage could then devote his whole time to Yard Department duties, which included duties of the Station Security Officer as well as the Passive and Air Raid Defense Officer. In 1942, the Yard Officer was also appointed the Ration Officer, which included rationing gasoline, tires, new cars and shoes. In 1943, additional duty as Station First Lieutenant was assigned.

The first boats were received by the station in September 1940, consisting of one forty-foot motor sailer and one motor whale boat. The present Boat House was then under construction and upon its completion in January 1941, the boats were moved there. The first crash boat, a twenty-nine foot boat, was received in January 1941. There were thirty small boats attached to the station besides the recreation sailing boats and the total crew in the Yard Craft Division numbered 171 men.

Within the Yard Craft Division was the Diving Section, consisting of a Diving Officer and six second class divers. Through the efforts of Ensign D. G. Ford, Diving Officer, the station obtained permission in 1944 to convert a fifty-foot motor launch in to a Diving Boat. The Diving Section was little known until it was needed, when it became an important and necessary part of water salvage. The section was ready at any hour and the men were trained to work together smoothly and efficiently.

Through the efforts of Admiral A. B. Cook, the first Chief of the Naval Air Operational Training Command (NAOTC), four ferry boats were obtained for the station in July 1942. Two of the boats had previously been used for carrying vehicles from Tampa to St. Petersburg so it was necessary to convert them to carrying passengers. The first trip from the Naval Air Station to downtown Jacksonville was made in August 1942. The boats were renamed the U.S.S. De Weese, 600 capacity, and U.S.S. Seabrook, 400 capacity. Two other ferry boats were obtained and used to transport passengers to and from the Bolles School Landing; the U.S.S. Caladesi, 200 capacity and the Tropic Star, 115 capacity.

The Seaman Guard was organized December 10, 1941, under the command of Lieutenant F. C. Suratt, USNR, with a complement of a small armory. Its purpose was to maintain internal security and to serve as a school for sentries. The men were assigned from various squadrons and departments on the station to serve a three-month tour of guard duty. Twenty of the men were assigned to duty at the Main Gate to serve as armed escorts for personnel and material coming onto the station.

As the station grew, the Seaman Guard under Lieutenant Suratt's able leadership kept pace by adding new posts and special assignments to its list of duties. A school for sentries was established, emergency squads were formed, escorts were provided for military functions and funerals, personnel were provided for parades and bond rallies, search parties were provided for lost personnel and crashed airplanes, a unit for fighting brush fires was organized, auxiliary shore patrolmen were furnished for bus and ferry patrol duties, a roving patrol of the unrestricted areas of the station was organized, and men were provided to assist in traffic control. In late December 1943, the Seaman Guard began to act as the U.S. Mail Distribution Center of the Yard Department.

Along with the increase in duties came an increase in mobility and firepower as a jeep, two canopy trucks, and a Fire Truck for brush fires were added. The armory was expanded and stocked with riot guns, automatic rifles, sub-machine guns, machine guns, Springfield rifles, training rifles and side arms in adequate numbers. In June 1942, the Seaman Guard received a shipment of sentry dogs from Front Royal, Virginia, thus becoming the second Naval Air Station in the United States to use sentry dogs. The Seaman Guard personnel were quickly trained in the handling of these dogs and within a few days were assigned to six of the vital posts. The Sentry Dog Unit served well and faithfully until January 19, 1944, when it was replaced by a Detachment of Coast Guard men and dogs. The men of the Coast Guard stood all watches on posts where sentry dogs were assigned and performed other duties as assigned by the Officer in Charge of the Seaman Guard.

The Seaman Guard, realizing the necessity for and the value of physical training, began early in 1942 to enter into the station's athletic program and was the first unit on the station to be assigned a physical training instructor. Seaman Guard personnel received thorough instructions in judo, hand to hand combat, wrestling, swimming, boxing and obstacle course conditioning. In addition, the Seaman Guard participated in organized station athletics; winning championships in touch football, water polo and baseball, and at the same time, piling up good records in bowling, softball and basketball.

In December 1940, the first piece of fire apparatus, a 500 gallon per minute pumper, was received. Since there were no fire houses at the time, it was parked under the shed of the Boatswain's Locker. The first fire that occurred on the station after this time was in a small contractor's shack located on Langley Street directly in front of what is now the Administration Building. In 1942, it became apparent that with the receipt of additional fire-fighting equipment and with the enormous buildup of the station, it was necessary to have another fire station located in the NATTC area, then called the Trade School. Fire House No.2 was built and equipped at the corner of Child Street and Enterprise Avenue. Later in 1942, it became evident that it would be necessary to have still another fire house, so Fire House No. 3 was built on Mustin Road opposite the Senior Bachelor Officer's Quarters. The station had three well equipped and well manned fire houses, consisting of four 750 gallon per minute pumpers, one combination ladder and pumper, five Bean H. P. pumpers, and two C02 pumpers for crashes, and there were 124 men, a Chief, and an Assistant Fire Chief in the Fire Department.

When the station was first commissioned on October 15,1940, there was no fence around the station, and all traffic used what is now the north gate. There were no paved roads and all traffic going south used the state highway, or what is now Allegheny road. Upon completion of the fence and the moving of the new state highway to outside the station in March 1941, a Marine Guard was established and passes were required of all personnel coming on the station. In December 1940, the first station automobile tags were issued, and all service personnel were required to have one to bring their automobile aboard the station. The number of station passes that were issued increased from a few hundred in 1941 to nearly 30,000 in 1944. The Yard Office issued all permanent station passes, except for enlisted personnel, and a complete file of all applications with fingerprints was maintained. As a convenience to station personnel, state automobile tags, drivers' licenses, fishing and hunting licenses were also sold in the Yard Office. In June 1944 the Gasoline Rationing function was moved to a separate office to be managed by personnel in Building 450, and the Yard Office handled only gasoline for authorized travel between posts of duty on station. The first station Chief of Police and Chief Master-at-Arms was Chief Boatswain's Mate C. Hawk. He and two Boatswain's Mates comprised the entire Police Force but with the growth of the station, additional men were necessary, and in 1945 there were fifteen men in the Police Department. With miles of roads and several thousand automobiles on the station every day, the station police had the same problems as a city police department, and in addition they had to handle all mast cases and other Navy duties. In the latter part of 1940, the Safety Division was organized and in the capable hands of Lieutenant Commander C. O. Bain, who was assisted by Robert F. Taylor, acting in the capacity of Safety-Inspector, and one clerk typist. Due to the tremendous amount of construction work in progress, hazards developed that could not be coped with, and it was decided to secure the services of an experienced Safety Engineer. Mr. Eben P. Armstrong was brought in from California to set up a safety program that would reduce accidents of all types. This program was continued and expanded with excellent results. The station never finished lower than third in accident standing among air stations in the 1940s. The following also served as Safety Officer through the war years: Commander Windsor H. Cushing, Lieutenant Commander Walter B. Snook, Lieutenant J. A. Smith, and Lieutenant Moses F. Brinson. These officers made it possible for the station to establish a safety record that was the envy of Naval Air Stations throughout the country - there was not one fatal accident to a civil service employee on the station in the 1940s.

PUBLIC WORKS

DEPARTMENT

When Commander C. II. Cotter arrived on September 15, 1939, the Public Works Department in effect was born. Construction supervision was the main task until 1941, when a department was officially organized. By June 1941, there were five main sections consisting of Administration and Clerical, Maintenance and Power,

Transportation, Design and Engineering, and Housing and Real Estate. Reflecting the steadily increasing workload, the number of civilian employees in the Public Works Department increased rapidly. There were approximately 400 in March 1941, 700 in July 1941, 900 in December 1941, 1,070 in February 1942, 1,500 in February 1943 and 1,730 in April 1943.

Due to the growth of the auxiliary stations, it was considered advisable to establish Lee Field, Cecil Field, Naval Air Gunners School, and the Mayport Auxiliary Air Facility as separate commands under the Commanding Officer, Naval Air Station Jacksonville. Accordingly, on February 1, 1943, the Naval Air Gunners School, Naval Auxiliary Air Station, Green Cove Springs and the Naval Auxiliary Air Station, Cecil Field were officially established as separate commands. In 1944, the Naval Auxiliary Air Station, Mayport, and the Naval Auxiliary Air Station, Jacksonville Municipal #1 were established as separate commands. Independent public works departments were organized at each of these commands for maintenance and minor repairs, with construction and planning remaining under the Public Works Officer at NAS Jacksonville. A nucleus of trained civilian personnel was transferred to these facilities from the Public Works Department, NAS Jacksonville. The Transportation Section of the Public Works Department was established as a separate division under the Executive Officer early in 1943, with a corresponding transfer of personnel from the Public Works Department.

The positions of Public Works Officer and Officer-in-Charge of construction were both occupied by the same officer throughout the war years. The two combined offices in 1943 developed into seven distinct sections, with functions as follows;

(Clerical Administrative Section - Clerical, administration, employment, dissemination of work.

(Contract Section - Had cognizance overall construction by contract including inspection, contract changes and interpretations.

(Maintenance Section - Operation of various shops for maintenance, alterations, repairs and minor construction of station facilities.

(Utilities Section-Operation of steam generating plants, water and sewage treatment plants, sewage pumping stations, swimming pool

equipment, water purification systems, emergency electric generators and cold storage plants.

(Design and Engineering Section - Planning and designing of new station facilities, alterations and repairs.

(Housing and Real Estate Section - Management of four Navy-owned housing projects consisting of 600 family dwelling units, acquisition of land, matters relating to access and closure of roads.

(Chemical Laboratory Analyses in connection with purification and sanitation of water and sewage conditioning of boiler water and pipe corrosion.

The Public Works Department came up with many innovative ideas during the war years to save scarce materials. In 1943, Maintenance Section personnel procured a second hand sawmill, and as areas were cleared, boards were made. They also built their own dredge for dredging and filling operations along the seawall.

The Maintenance Section, with 1,200 employees, was responsible for all base maintenance, including the operation of the station railroad. The entire department found innovative ways to solve some extremely difficult wartime emergencies and problems and the actual commissioning date of the station was six months sooner than expected due to the leadership of the first Public Works Officer.

MEDICAL DEPARTMENT

The first representative of the Medical Department, Chief Pharmacist Mate A. E. Owen, reported to the Officer in Charge November 11, 1939. A temporary dispensary was set up in the former Camp Foster Administrative Building (The Facilities and Environmental Department building today.) On March 15, 1941, Commander E. E. Smith reported and assumed the duties of the Medical Officer

Commander Smith was promoted to Captain in July, 1941, and served as the Senior Medical Officer for most of the war years.

The main dispensary building, building 8, was made of permanent concrete construction, different from a lot of construction at the time. Work began on the building in October 1940, and it was occupied in March 1941. In July 1942, the work load on the main dispensary became so heavy that a barracks across the street from the dispensary was assigned to the Medical Department as a dispensary annex. Construction of an addition to house the refrigerated low-pressure chamber and accessory activities began in October 1942.

The station dispensary was already inadequate for station needs at the time it was commissioned. The medical officer made repeated efforts to obtain enlarged facilities and eventually construction of a major addition to the main dispensary began in April 1943.

Complete facilities for the examination and treatment of all personnel were finally provided. Those requiring major surgical attention or seriously injured or ill patients were transferred to the Naval Hospital.

YELLOW WATER NAVAL AIR GUNNERS' SCHOOL

The Jacksonville Naval Air Gunners' School at Yellow Water was the largest facility of its kind in the Navy. Its mission was "to supply a constant and orderly flow of fully qualified combat air crewmen to the Fleet."

Naval Air Gunners' School was authorized on April 2, 1942, construction began in May 1942, and the school was commissioned on August 28, 1942.

Starting with some eight officers and seventy odd enlisted men, the school grew to more than 3,000 enlisted men and 105 officers by the end of the war. In the three years of its existence, more than 30,000 gunners had been trained for duty in the Navy's huge patrol bombers, torpedo bombers and dive-bombers.

Fourteen square miles of swamp land, fit only for cattle-grazing and turpentine ranching, had been transformed into the settlement with miles of paved streets and sidewalks, sewage disposals, electric, steam and water plants.

Eight miles of railroad tracks were used by the target cars and an astronomical number of rounds of ammunition were fired and not one person was seriously injured in live fire training!

PBYs on the seawall, 1945. (U.S. Navy)

THE 1940s

1940

The site for NAS Jacksonville was mostly consumed by development and initial construction activities in 1940. There was a tremendous task at hand in laying out the station, and by looking at the station today, one that was accomplished with careful forethought. The buildings and roads were all laid out and were to be numbered logically. The buildings in the 500 series, for example, were all placed on one block. The 600 series buildings were placed in another block, and so on. NAS Jacksonville has lost this system of building numbering today, and buildings are now numbered in a more haphazard manner. Not only was a new base to be built, but the final remains of Camp Foster were removed. Only three original buildings were left when construction started, and only one still exists today (Building 27, where the Facilities and Environmental Department has offices).

The first real highlight in January 1940, was the arrival of the first aircraft to be assigned to the base, a Grumman Duck, which would eventually be Captain Mason's private plane. It was kept at the Municipal airport and flown in from San Diego by Commander Jimmy Grant, who would be the station's first Executive Officer. The most visible items of Camp Foster were moved on January 18, as the cannons, which were located at the main entrance, were relocated to Camp Blanding. The cannons were captured by American forces from German and Austrian Divisions in World War 1. An Act of Congress in 1924 allowed them to be donated to the State of Florida, and they were placed at the National Guard Camp. Two Austrian 88 mm guns were moved first from the main gate, a 210 mm gun was moved from the Camp Foster Administration building, and a 250 mm gun was moved from the parade ground area. All are on display today at Camp Blanding.

On January 19, construction started on the Overhaul Shops, which would be the nation's largest. In addition, work started on the landing mat and pier. Commander Grant gave the first picture of what the base would look like at a speech before an advertising club in Jacksonville. Not only was 30 percent of the total Navy allocated construction funds going to NAS Jacksonville's development, but NAS Banana River would also be designed and built by station Public Works personnel. (Banana River today is Patrick Air Force Base near Cape Kennedy).

On April 8, plans were being developed and funds procured for a new highway to the base. Construction would not start until October 25 and it would eventually be called Roosevelt Boulevard.

May 1 saw the first Marines arrive to start guarding the property. They were initially put in a two-story brick building left over from when Yukon was part of the station. It had been Crosby's Store. The old two-story brick store was also converted into Marine Barracks. This area was located where the station's commercial gate is today, at the intersection of Albemarle and Allegheny. The Marines mostly handled traffic and protected property against theft. It wasn't until 32 more Marines arrived on July 1 that tightened restrictions for access into the station went into effect. The Marines received their first motorcycle on July 7, and a second went for messenger services.

A letter to Captain Mason (future first Commanding Officer) from Commander Cotton (who would head the Trade Schools) was already giving him suggestions for improvements. He stated "While in Norfolk I inspected their auditorium… There are some glaring defects in the one at Norfolk that were pointed out to me, and they should not be repeated in those in Jacksonville."

In September the station runways had progressed to the point where the first plane could land there. The Marines were now also providing full security as they were up to 150 men. So choice was the assignment to Jacksonville that two marines actually paid their own expenses to get here!

A big event for the station's chief boatswain was the arrival of his first two boats on September 9. The Florida Times Union on September 10 stated he "was handing out cigars faster than a father of twins this morning and almost for the same reason." The boats would serve as utility and crash boats. Captain Mason finally arrived in Jacksonville on September 17 and set up in a room at the Hotel George Washington until his on-base quarters were ready. Work on the dispensary started the last week of September. One of the unique features to be included was a special low pressure chamber to simulate flight training for high altitudes. The building had to be ready for use by December 1.

On October 2, Undersecretary of the Navy James Forrestal made a tour of the progress. Funds allocated for construction had been increased to $21,000,000. The official commissioning date of October 15 was formally announced on October 4 by Captain Mason. Only six other officers were now permanently based at the station. They were Commander Carl H. Cotter, Public Works Officer; Commander V.F. Grant, Executive Officer; Commander J.W. Vann, Senior Medical Officer; Commander Junious L. Cotton, Head of the Trade School; Commander T. Earle Hipp, Supply Officer and Major Erwin Mehilenger, Commanding Officer of the Marine Detachment. Six different hangars were in full construction during this period, as well as other construction activities over the entire station.

The station was commissioned at noon on October 15. One of the prominent visitors was Captain A.C. Read, Commanding Officer of NAS Pensacola. Captain Read was famous as commander of the first successful flight across the Atlantic.

The first nine officers' homes were finished before the ceremony so some could now move permanently on-station. Not only was construction proceeding at NAS Jacksonville, but facilities were also being constructed at Green Cove Springs for an outlying field there. The next day after the ceremony, Commander Hipps, Supply Officer, became the first person to move into his permanent offices and start work, as his building, 110, was completed.

The first aviator to report for duty, Ensign John E. Muldrow, arrived from Coco Solo, Canal Zone on October 21. Some 47 more would arrive in the next few days.

In November the station had its

first two fires. One caused about $6,000 worth of damage to electrical equipment and the other did $500 in a building in the cadet training area. The Marine Guards did a good job in keeping the fires from spreading to adjacent buildings. There was no pressure in the installed water mains to fight the fires, but that was quickly fixed the next day! On November 14, two more officers arrived for duty. Lieutenant Commander R. L. Bowman, who would be superintendent of the training schools and Lieutenant Commander J. B. Dunn, who would be senior seaplane squadron commander. The first 400 enlisted men were to arrive by December 1 as the barracks and galley were completed. Commander Cotter, who had supervised construction since the beginning, was the first officer to leave on November 26. He was transferred to Norfolk and Commander Meade took over duties for all station construction. In addition, he was responsible for construction at Lee Field, NAS Banana River, the Miami Naval Air Station and a reserve center in Atlanta.

December 4 saw a visit by Navy Secretary Frank Knox. He went to Captain Mason's office and got a detailed brief of station activities. Three more planes arrived on December 10, making for a total of four! Two would be used for utility purposes and the other, an aging Grumman fighter, would be disassembled and used for instruction purposes, never to fly again. The first 25 cadets arrived on December 30. Three would eventually become the first cadets to receive their wings at NAS Jacksonville. By December 30, the station also had 25 new N2S Stearman training planes. The first nine arrived on December 24.

1941

The year 1941 had to be one of the most dynamic in the station's 60-year history. Construction was being conducted all over the station; squadrons were to be formed; the Overhaul and Repair shops were starting to rework aircraft; a Naval Hospital was to be built; and every day new Sailors and Marines were being assigned to the station. The construction alone was mind numbing for the station Public Works Officer and the staff he was still building! Not enough credit can be given to these early station pioneers dealing with the sheer magnitude of their tasks at hand.

The first item to take place in 1941 was to stage a media event for the first flights. On January 2, 25 new N2S Stearmans were lined up for a flight in front of the local media. The instructor pilots were ready, as were 25 new students who would sit in the back seat of the two-seat plane. As the flight was ready to commence, one of the flight instructors asked for their flight goggles. Much to the chagrin of embarrassed station officials, they had forgotten to get them. A quick trip was made to Supply, only to discover there were not any in stock yet. The impressed fund "kitty" was used to obtain money, and a trip was quickly made to a local motorcycle shop to obtain goggles. When they returned, the first flights went on in front of media cameras!

January 23 saw the station's first construction fatality as a painter fell to his death. He had been painting the structural steel high inside Hangar 114 when a board he was sliding from one girder to another slipped.

By February 1941, Undersecretary of the Navy James F. Forrestal was paying an informal visit to the station he supported so much. Forrestal was still stinging from a recent decision to add another station at Corpus Christi. He was convinced NAS Jacksonville would be all that was needed to handle the training mission, but the political process was imposed to add the station in Texas. His February trip was also most probably an advance trip to prepare the base for President Roosevelt's visit a few weeks later. Just as Undersecretary Forrestal was reviewing the station, the first seaplanes were arriving. Some old P2Y-3's were brought over from Pensacola. The planes were obsolete, but would do for seaplane training until the PBY's arrived. Nine planes were brought to the station. Of note is when the PBY's arrived starting in June, no one wanted the old P2Y's. So a hole was dug near what is the warehouse area along Roosevelt Boulevard today and the planes, which were constructed of wood and canvas, were just buried. They remain there today! The last P2Y ended station service around 20 July. President Roosevelt's train came directly onto the station and stopped in front of one of the hangars promptly at 8:00 a.m. He took an hour tour of the station, and stated "I am very much surprised to learn it is as far along as it is." A new highway was being built to the station, and during his administration it was named "Roosevelt Boulevard".

Benjamin Lee II Field at Green Cove Springs was dedicated March 12 and on April 1, construction began at Air Auxiliary Station Number 2, which would later be known as Cecil Field. Cecil Field was originally only 2,600 acres and was bought for $16,851. Lee Field was the home of the first land based bomber squadrons, and when pre-operational phases were started it was the home of squadron VN-16, the largest of its type in the nation. This field was needed to provide additional space to NAS Jacksonville. The training mission, with planes assigned, was quickly over-running the present facilities. Lee Field was named after Jacksonville resident Benjamin Lee, the only aviator from Jacksonville killed in WWI.

Ensign George Gay reported to the station on April 3. He would go on to be famous because of the Battle of Midway. Ensign Gay was shot down on the first day of the battle and rescued the next. He floated among his plane debris and observed the entire battle close-up and later was able to provide vivid, first-hand observations and descriptions. On April 8, Raymond Allen, a former contractor employee, was arrested for taking pictures at the air station. Shortly thereafter, Captain Mason banned all cameras from the station - an order that lasted until 1946.

On May 26, the first cadet was killed while in flight training and Captain Mason later met the young cadet's parents at the airport. The young cadet was practicing low-level maneuver figure-eight turns in his N2S Stearman west of the station. He looped up, the plane rolled to one side and crashed. A dairyman, F. W. Ware, and his farmhands went to the field to see what had happened and discovered the pilot still alive. Ware's wife called the hospital and an ambulance was dispatched. He was transported to St. Vincent's Hospital and died that evening. Lieutenant Commander Joseph B. Dunn, officer in charge of Cecil Field, was appointed to investigate the mishap. Captain Mason and his wife met the young cadet's relatives at the Jacksonville Municipal Airport and escorted them to the base. He also attended all the services with the family including the burial at Oaklawn Cemetery. Unfortunately, as Cadet training fatalities increased to one, two and sometimes even more per month, the Commanding Officer could no longer be so personally involved in each case.

The first on-base wedding took place in the Chapel on June 21 and three days later the first graduating class received its wings. On June 28, NAS Jacksonville was on the NBC radio network with a live broadcast from the station. Rudy Mick was assigned to VN-15 at the time and had the following description: *"The day dawned with heavy overcast clouds that held the capacity to wash out everything. The script read that when Mr. George Hicks (NBC Announcer) arrived at our hangar, he would immediately board a waiting PBY. Once inside he would describe the aircraft, interview our own "Skinny" Wingfield, and depart the plane before it taxied down the ramp and eventually took off. I was to assist Mr. Hicks in boarding the plane, without talking in his microphone while in the plane and assist him in leaving the plane. When Mr. Hicks arrived we were in the middle of a torrential downpour. A PBY had been positioned with its nose almost against the hangar door, with the idea to start the engine on cue and simulate taxiing and taking off by revving up the engines. Needless to say chaos reigned and nothing followed the script. It took two attempts to get the first engine started but we were never able to start the other engine. Somewhere in the mounting pressure of trying to start that engine, either "Skinny" or I uttered "What the Hell!". Naturally it went out over the air as we were later told. So for the radio audience the taxiing and take off were accomplished on only one engine. No doubt a "first" for a PBY parked at a hangar."* Since VN-15 was the last stop on Mr. Hick's tour, Mr. Mick asked for and received the copy of the radio script for that broadcast. Mr. Hicks, when asked if he would autograph it, signed *"What the Hell! - What a Broadcast. George Hicks, NBC 41"*. Mr. Mick has treasured this script ever since.

The day before this NBC broadcast, the station had just commissioned the first Cadets. Ensign Clifford Hemphill, Ensign Jefferson Kennedy, and Ensign George Shortlidge were the beginning of a long line of future naval aviators.

The new hospital was commissioned in July, and a lightning storm, which hit right before ceremonies commenced, blew out a fuse, almost canceling out the ceremonies. The first patient reported the next day, suffering from an ulcer! Just one week after commissioning, Chief Petty Officer William Francis Swift died of a heart attack, becoming the new hospital's first fatality. Also the Navy received title to Ribault Bay (now Mayport), for a purchase price of $28,000. This property was paid for out of the $1.1 million bond issue passed in 1939.

On Saturday, September 5, at 8:30 A.M., Captain Mason called all station personnel to the drill field. This was the first time that all personnel had been assembled in one location. Captain Mason proceeded to thank the personnel for all their efforts in helping to accomplish the station's remarkable achievements in such a short time period. Mr. A. L. Gates and Admiral John Towers arrived for a visit on September 22, and the station football team, "Fliers," started with former Yale football players in October. 1941 was a year of many "firsts" for the station, but one which station officials could have done without was the first murder. On November 8, two mess attendants strangled another mess attendant with hammock lashings with robbery as the motive, wrapped chains around his body and threw it into the St. Johns River. The body was found shortly thereafter at the pier behind the quarters where the present Admiral's quarters are today.

December 7 brought new urgency to the station's mission. As the news of the Japanese attack at Pearl Harbor hit station personnel, anger was the first emotion felt! Machine guns were immediately placed on the station air operations tower, and armed guards were placed inside every hangar. One of the aviators killed at Pearl Harbor received his wings at the air station in the first graduating class. Sailors out in town were called back to the station, many having a dance cut short in Jacksonville. Aircraft were moved to Cecil Field and training flights started there the very next day. At the reunion of the early squadron pioneers in 1995, Leo Jenson recounted the story of the first night with aircraft at Cecil Field. "We had moved aircraft over from NAS Jacksonville, mostly Stearmans, N3N's and some NR-1's. The landing area of Cecil was still a cow pasture that had not been totally fenced off from the cows yet. A watch was stood up and a guard, who's name I can't remember today, but he was from New York, got the first watch. At around midnight, the guard came running back where we were sleeping yelling 'The cows are eating the planes, the cows are eating the planes!' We told him to go back and fire his rifle and chase them off. The next morning we discovered the cows had eaten the ailerons off approximately 40 aircraft! Guess they liked the glue and fabric around the wings". NAS Jacksonville, and the men at Cecil Field, were now preparing for war.

The First Offensive Action of WWII?

Over the last ten years, I probably have gotten close to 40 letters from Leonard Kiesel, U. S. Marine Corp (Retired). Mr. Kiesel was stationed at NAS from practically the beginning of the stations history in 1940 through early 1942. He remembers a three-alarm fire that broke out in Jacksonville's warehouses near the docks, one that required additional NAS Jacksonville assistance to fight. He and some other marines assisted. He also remembers the fire being in an area where liquor was stored, and the marines "rescued" 50 cases of liquor only to be later consumed at a tavern near the NAS Jacksonville front gate! His real story, however, involves an action that even today has considerable circumstantial evidence, but no actual written record. It involves possibly the first offensive action against an enemy in the United States for WWII.

On Sunday, December 7, 1941, the news of the attack on Pearl Harbor was just starting to go through Jacksonville. Apparently in the Jacksonville port were three Italian freighters, the *Confidenza*, the *Ircania* and the *Villarperose*. The ships hearing of the attack on Pearl Harbor apparently got spooked and headed for open water. Then NAS Jacksonville Commanding Officer, CAPT Charles P. Mason, ordered the 30 Marines in the security force into three barges with instructions to capture the crews. The Marines quickly overtook the unarmed vessels, climbing ropes and cargo nets to board them. None of the Italians spoke English, but the Marines with bayonets fixed had no problem communicating what they wanted done. The Italians were loaded into the barges and taken back to NAS Jacksonville. When Mr. Kiesel left the station in February 1942, the Italians were still being kept at the station in a barracks building converted into a makeshift guardhouse.

My research into this found no evidence in the local Jacksonville papers or station historical records, but that would not be surprising. The Navy had control of stories that went into the paper during WWII, as noted in the stations public information office statements in historical records. Also, many events considered classified that took place at NAS Jacksonville in WWII were not documented in files available even today. Secret testing on German parachutes was done at the station in WWII, but no written record of this event exists either. Only recently did a photograph come to light showing this event. Mr. Kiesel, and the four remaining Marines alive today who also took part in this event, have been fighting to get the taking of these Italian ships officially recognized ever since, with not much luck. In 1995 I had an old gentlemen enter my office to "talk to the base historian", and he relayed me his story of having to guard Italian prisoners at the start of 1942. He never understood where they came from or why they were at the station. He had also never heard of Mr. Kiesel when I asked so I gave him his address and told him to write him! The only other piece of information I have ever found was in the book *The Jacksonville Story 1901-1951*, published by Jacksonville's Fifty Years of Progress Association in 1950. On page 113 under the year 1941, it notes "Two Italian ships seized."

The First Flight Class

Flight classes were formed to teach students how to fly beginning in December 1940. Class 12A-40-J was the first formed and the designation had the following meaning applicable to all future classes: 12 was for the month formed (December in this case) and each class formed that month was also designated by letter starting with "A"; 40 was the last two digits of the year formed (1940); and "J" was for Jacksonville. This first class received their wings on June 27, 1941 with the first three (Ensigns Hemphill, Kennedy and Shortlidge) receiving their wings on June 24. Of this first class of 27 new aviators, only 5 were killed during WWII. Ensign Mears was killed in a flight training accident at NAS Miami; Ensign Cassidy was killed during the attack on Pearl Harbor; Ensign C.M. Kelly would be lost at sea during the Battle of Midway; Ensign H. R. Keller was lost at sea in the Pacific; and Ensign N. Seiller was killed in a plane crash in the North Pacific in January 1945 while attached to the USS Langley. In 1990, 16 of the original 27 made it back to the station for the 50th Anniversary Ceremonies. *(Photo next page.)*

1942

1942 continued with the same rapid expansion as 1941, but with a renewed sense of urgency. Construction of buildings was being completed and turned over to the government almost weekly. In the first week of January, two cadets were killed in flight training and three more died just a week later. This primary training was to be eventually phased out, though, as advanced training was to be phased in.

The war came close to Jacksonville on the night of April 10, 1942. A German U-boat torpedoed the tanker SS Gulfamerica off the coast of Jacksonville Beach. The station immediately sent PBY's to investigate and destroy the submarine. The submarine escaped, but there was a new sense of urgency with the PBY training. The next month VP-83 was assigned to the station for coastal patrol duties. They were relieved by VP-94 a month later. The antisubmarine patrols continued through August. On April 14, a PBY "Catalina" crashed near the Shands Bridge at Green Cove Springs, killing two personnel.

The Naval Air Operational Training Command was established with headquarters at NAS Jacksonville on April 30. The primary purpose of this command was to relieve the fleet units of the responsibility of training pilots, consequently enabling the Fleet to direct all efforts toward combat operations. The secondary purpose was to coordinate all phases of operational training under an officer of Naval Aviation. One change for the Commanding Officer of NAS Jacksonville was he lost his residence on base to the new Admiral. It has been this way ever since.

On May 1, America's number one air ace, Navy Pilot Eddie O'Hare, visited the station. O'Hare, for whom O'Hare airport in Chicago is named, had just received the Congressional Medal of Honor for shooting down five Japanese planes and damaging a sixth in one afternoon. It was the first Medal of Honor personally awarded by President Roosevelt. He gave a talk to station officers and cadets on the enemy they might confront. Later in May, one of the largest air armadas ever assembled in Jacksonville took to the air. A total of 254 planes flew overhead on May 6, to impress Pan American officials. The viewers were awed at the spectacle and inspired by the number of warplanes parked throughout the station's flightlines. On the previous day, Joseph P. Kennedy received his wings at the station. His father, who had been Ambassador to Great Britain, was on hand for the event.

Captain Mason departed in May and Captain John Dale Price became the station's second Commanding Officer. On May 7, the flag of Rear Admiral Arthur B. Cook first flew over Building 1 as he assumed command of the new Air Operational Training Command. Along with this came more advanced fighter aircraft to the station. As ceremonies were taking place, Lt. Commander Gage of the Yard Department was warning the citizens of Jacksonville about removing the sea plane drone lights that marked landing sites in the St. John's River. "These buoys are not sold to civilians," said Commander Gage and "anyone found with any of them in his possession is liable to a heavy fine and imprisonment. We will see to it that they are vigorously prosecuted."

Later in the month, one of the more spectacular crashes occurred on May 27, when a plane crashed in front of the base Dispensary, witnessed by hundreds of station personnel. Pharmacist's Mate Third Class M. B. Mulligan jumped into the flaming wreckage to rescue the pilot, who unfortunately died two days later. Mr. Mulligan himself was severely burned, requiring hospitalization for two and a half months. He was later awarded the Navy and Marine Corps Medal for his outstanding courage. Another midair crash over the airfield killed four more when two Stearmans collided head on July 13.

Since only one bridge was available at this time for civilians who lived in the Southside to commute to NAS, a ferry was finally procured in August to offer some relief. The "Seven Seas" started service on August 30, carrying passengers to the Seaboard Railway Dock at the foot of Hogan Street. Two other ferries, the "Manatee" and "Pinelias" were later procured from Tampa. Gasoline and tire shortages made the ferries a very popular form of transportation with base personnel.

The Aviation Free Gunnery School was commissioned August 30 at Cecil Field (Whitehouse). As Commanding Officer Price officiated, students were already in training. A machine gun, handgun and shotgun range were all established at this time.

Another former member of the station, Commander James H. Flatley, Jr., was awarded the Navy Cross for his heroism in the Battle of Coral Sea in September. He was credited with shooting down five Japanese planes while trying to protect the Navy carrier USS Lexington. When the second

FLIGHT CLASS

12 A - 40 - J

John Carter
East Orange, N. J.
Duke University

Arthur Cassidy
New York City, N. Y.
Fordham University

Bernard Gallagher	Clifford Hemphill	Foster Kay	Harold Keller, Jr.	Robert Keller
Hartford, Conn.	Spring Lake, N. J.	Hartford, Conn.	West Newton, Mass.	Berkeley, Calif.
Georgetown Univ.	University of Virginia	Mass. State College	Williams College	Univ. of California

Justin Miller	Frank Mears	Douglas Mulcahy	William Murray	Robert Nelson
Baltimore, Md.	Pocomoke City, Md.	Yonkers, N. Y.	Philadelphia, Penn.	Frankfort, N. Y.
American University	Maryland University	Carnegie Inst. Tech.	Catholic University	Clarkson College

Arthur Seaver	Edwin Seiler	Ernest Sexton	George S' ortledge	David Solomon
Manhasset Bch, N. Y.	Greenbrook, N. J.	Wytheville, Va.	Chesham, N. H.	Paterson, N. J.
Hamilton College	Princeton University	Bluefield College	Harvard University	New Jersey S. T. C.

George Crittenden
Mineola, N. Y.
Ohio University

Michael Elin
Hartford, Conn.
Virginia Poly.

C. M. Kelly, Jr.
Baltimore, Md.
Univ. of Maryland

J. Kennedy,
New Rochelle, N. Y.
Rollins College

Clarence Quinlan, Jr.
Warwick, R. I.
Rhode Island State C.

Thomas Seabrook
Ridgewood, N. J.
New York Univ.

Humphrey Tallman
Fairhaven, Mass.
St. Lawrence Univ.

Arnold White
Circleville, Ohio
Ohio Wesleyan

Louis Wilds
Aiken, S. C.
University of Virginia

Charles Willis, Jr.
Baltimore, Md.
University of Florida

anniversary occurred on October 15, there was little fanfare. Captain Mason had just recently been recommended for the rank of Rear Admiral and Captain Price was busy "supplying vigorous leadership in an even greater air school." The last formal act of 1942 was commissioning Naval Air Station Lake City. The station would immediately transfer multi-propeller training there, using SNB's.

Vernon Howe Bailey's

Watercolors

In March 1942, Vernon Howe Bailey spent three weeks at NAS Jacksonville completing a series of 22 drawings and watercolors. Mr. Bailey had been commissioned by the Navy to go to several Naval installations to make an official record of activities through his drawings. He was impressed with the activities he saw while at the station, particularly with a plane taking off every few minutes.

Mr. Bailey had come to the station with a long history himself. He was the first artist authorized by the U. S. Government to make drawings of America's war effort in World War I. He did drawings on activities taking place at gun shops, munitions plants and navy yards. Those prints were shown in the French War Museum. He also had done a series of watercolors for the Vatican.

Mr. Bailey's time spent at NAS Jacksonville documented activities throughout the station, including aircraft rework at the Assembly and Repair Department, the band, activities sailors were performing such as parachute rigging and even a print of the Commanding Officers house. When he finished his series, all of the prints were presented to the Commanding Officer, Captain Mason. These prints provided a rich record of the early war efforts of the station. The original watercolors are catalogued today at the Washington Naval Yard in the section where a large cache of historical paintings are kept.

1943

1943 started out with a change of the name of Service Schools to the Naval Air Technical Training Center and with the dedication of the Catholic Chapel. Although services had already been held since December 1942,the formal

Spirit of '42—Call to chow. (Bailey #200)

Crane hoisting seaplane . (Bailey #199)

Land planes, NAS Jacksonville. (Bailey 200a)

Assembly and Repair, Interior of Hangar 101. (Bailey 184a)

dedication had been delayed for various reasons until January 17.

The first contingent of enlisted WAVES (Woman's Auxiliary for Volunteer Emergency Service) reported to NAS Jacksonville straight from boot training in January. This group was made up largely of seamen, with a few storekeepers, yeomen, and communication workers. Some officers were already onboard, making preparations for the enlisted WAVES.

Within a month, a second group arrived and soon it was customary to welcome new WAVES onboard every time Hunter, the new WAVES Training School, released a new class. Before the end of the war, more than 1,500 WAVES, enlisted and officers, were on duty here.

In the early days, WAVES were assigned to do desk work only. But as new needs arose, new skills were developed and by 1944 they were wearing distinctive sleeve badges testifying to work in 38 different ratings. They proved they were capable of holding down billets as aerographers, mechanics, radiomen, electricians, printers, cooks, parachute riggers, and photographers. They taught navigation and gunnery and conducted simulated flights in the Link Trainer.

As the Naval Air Technical Training Center's second birthday approached on February 15, it was recognized as "one of the largest in the world" and as "a big part of the arsenal of democracy." Numerous schools, teaching double shifts to both Navy and Marine personnel, had developed at the center since its meager beginnings just two short years ago.

On February 20, both Lee Field in Green Cove Springs and Cecil Field were changed in status from outlying fields to auxiliary Naval Air Stations.

Both stations were started when training planes, cadets and instructors were sent from NAS Jacksonville to commence training. Both stations were also busy with construction of facilities in order to keep up with the additional squadrons assigned and the nation's ever-growing war effort. Cecil Field was named in memory of Commander Henry Barton Cecil, who was killed in April 1933 when the U.S.S. Akron, a dirigible, crashed off Barnegat Light, New Jersey. Cecil had been a friend of Captain Mason, the first Commanding Officer of NAS Jacksonville. He laid the groundwork for naming the new station in memory of his friend.

On February 26, after providing training to some 2,200 cadets, the station's last primary training class of 10 pilots graduated. Concurrently, the days of seeing aircraft such as the N2S Stearman, N3N and NR-1 training planes in the Jacksonville skies came to an end. This mission was transferred to NAS Corpus Christi. As this was ending, training started in the OS2U "Kingfisher" to include catapult practice. A catapult was added to the station pier so the plane could be launched over the St John's River. Gerald Lahay, stationed in the squadron, remembered a catapult launch that got him into trouble. He shared it with me as I gave him a tour of station facilities and he saw the pier in 1995. "There was a line of guys waiting to be launched. A crew would get in and the plane would launch. After a short flight after being launched, you would have to land in the river, taxi back to the station and the plane would be placed back on the catapult for the next crew. It took approximately 30 minutes between launches. As I got in the plane, the officer in charge told me to make it fast, as folks were waiting their turn. They launched me, I went about 100 feet, landed in the river and immediately taxied up the seaplane ramp. I got out, walked down the pier and tapped the officer on the shoulder, who was busy talking and never saw my launch or landing. I asked him if that was fast enough, (taking all of two minutes) and boy did I catch hell."

The Kingfisher pilots were not the only ones with a sense of humor, however. Seems you were not initiated in PBY's at NAS Jacksonville until you had fallen asleep on a flight and someone would put a match between your toes and light it!

The All Saints Chapel was dedicated on March 1 and the station's third Commanding Officer, Captain Stanley J. Michael, relieved Captain Price. Captain Price received orders to sea duty and like Captain Mason before him, a promotion to the rank of Rear Admiral. March 6 saw a freak storm hit Cecil Field. The wind started kicking up fast and windows started shattering in Hangar 13. Men started running for cover. When the storm ended as quickly as it had begun, over 100 of the stations aircraft were scattered throughout the station and surrounding woods. Most were badly damaged. The station ready plane was found in Orange Park when the resident called NAS Jacksonville to report an aircraft in her back yard! No reports of the damage were provided to the local citizens of Jacksonville.

In April, the film "Air Crew" was being shown in the Florida Theatre. This film was significant in that it was shot almost entirely at NAS Jacksonville. On Saturday, April 17, an impressive ceremony was held at the station bandshell. Major Alexander Brest, a longtime Jacksonville resident for whom the Brest Planetarium at the Museum of Science and History was named, accepted, on behalf of contractors, an Army-Navy Production award for his work on Navy projects. Along with Brest, Auchter Construction, Duval Engineering and Baston-Cook Company received awards. What accomplishments had they made? This team built more than 700 projects at NAS Jacksonville including the Naval Air Technical Training Center, the Naval Hospital, NAS Banana River, NAAS Lee, NAAS Cecil Field, the Gunner's School, the Dry Docks in Jacksonville, the facilities at Mayport, the Coast Guard Station in St. Augustine, the Recruiting Station in Jacksonville, and the landing fields at Lake George, St. Augustine, Switzerland, Fleming Island, Whitehouse, Branan, Herlong and many other Auxiliary Landing Fields. Mr. Brest personally presented me with his program from this event shortly before he died, still very proud of this accomplishment. The total costs attributed to this one large contract exceeded $38 million. Later in April, the post office established box numbers to the station's 30 activities, many of which have the same numbers today.

The Aviation Supply Officer School graduated its first class in May as the Air Operational Training Command was celebrating its first anniversary. Rear Admiral Mason returned to speak at the station "bandshell" on May 28, and told of his experiences while skipper of the ill-fated aircraft carrier "Hornet" which was sunk in its final battle with Japanese forces on October 26, 1942. Rear Admiral Mason received the Navy Cross for his outstanding courage and heroism while defending the "Hornet". Of the 2900 personnel assigned to the ship, all but 120 were rescued.

After only six months of command, Captain Michael, who had been promoted to Rear Admiral, was detached for sea duty and Captain Arthur Gavin become the fourth Commanding Officer on August 7, 1943. At that time, NAS Jacksonville was among the three largest naval air stations in the world and known as the "graduate school of naval aviation." Shortly thereafter, Rear Admiral Andrew McFall succeeded Vice Admiral Cook as Chief of Naval Air Operational Training.

In September, the first female Marines reported for assignment as storekeepers in the Assembly and Repair Department. Captain Gavin also imposed an 11:00 P.M. curfew for enlisted personnel.

October saw fighter training commence at the station with the arrival of Corsairs and Hellcats. NAS Jacksonville was becoming the home of advanced fighter training for the Navy. Actual operational training in the Corsair started on October 26 and by November 12, the torpedo-bomber training was moved to NAAS Cecil Field. Squadron VF-5 was established November 18, strictly for Corsair training.

In November, flight training personnel received talks from Lt. Kenneth Walsh, USMC, America's number two war ace with 20 Japanese downed planes to his credit. Major Joe Foss, America's number one war ace, had already spoken to the station's aviators some two months earlier. His record did not last long though, as Major Greg "Pappy" Boyington of the famous "Boyington's Blacksheep" soon tied it.

December 18 saw another crash, which was the talk of the station. A PBY "Catalina" crashed where the entrance to Bayview Elementary School is located today on Lake Shore. Mr. Billings, who lived on Cedar Creek at the time and saw the plane crash, ran and rescued five crewmembers. The station Commanding Officer later gave him a citation.

STATION INSIGNIA

Most station personnel today if asked what the official station insignia (or logo) is, will immediately tell you it is the triangle shaped logo seen throughout the station. And if you were to ask anyone if this has always been the station logo, most would probably also say yes. But, in fact, the present logo is the fourth the station has had since inception.

The first logo was actually a flying squirrel. On March 7, 1941, Aviation Cadet Clifford Hemphill, Jr., wrote to Walt Disney asking their help in designing and permission to use an insignia for the station. In his letter he stated: "By way of introduction let me state that I am an Aviation Cadet of Class 1 attending the new Naval Air Station here at Jacksonville and have been put in charge of the Art Department of the JAX Base portion of the *FLIGHT*

Station's first insignia.

Station's second insignia.

Station's third insignia.

Walt Disney designed Gunnery School Patch.

Station's insignia today (adopted 1963.).

JACKET, annual publication of Pensacola Training Station." (Note: This book would include half scenes and pilots from Pensacola and half from NAS Jacksonville.) "My problem is in designing the official squadron insignia and realizing that J. Gosling (Donald Duck) is the insignia and mascot of Pensacola. I thought you could help me." Cadet Hemphill went on to say he had an idea for an official insignia. He wanted a flying squirrel with some expression of semi-astounded surprise and happiness on its face. It would embody the elements of a novice at flying, an appropriate symbol for the stations' novice aviators. The insignia was designed and furnished by Walt Disney Productions back to Cadet Hemphill in a letter dated March 21. That was a very fast turn-around time for Disney, who was swamped with requests for insignia. This insignia was then adopted for the station's use almost immediately.

Just three months later, on May 8, 1941, Commanding Officer Captain Mason wrote another letter to Walt Disney. In that letter he stated; "I know that you have been plagued recently by requests from aircraft squadrons desiring insignias. Primary Training Squadron Eleven-A has been casting about for some time for an appropriate insignia, and has decided on the enclosure. Permission is requested to use this replica of Donald Duck's nephew as the official insignia of training Squadron Eleven-A." The letter was received on May 16 by the Walt Disney Productions and again, almost immediately drawn and sent back to the station. When Captain Mason received this insignia, he must have liked this design much more than the flying squirrel, as the squirrel all but disappeared almost immediately. The second insignia was used not only by

squadron VN 11-A but throughout the station until primary training was discontinued in 1943. Aviation Cadet Edward Skully actually designed this second insignia used by the station. He was the photographic editor of the Jacksonville section of the 1941 *Flight Jacket* year book.

The third insignia was requested from and approved by Walt Disney in June 1943. In a letter from the station, Disney was asked to design an insignia showing the baby duck having grown up, tossing his training book aside and showing determination to go after the enemy. The third insignia showed that very thing, even if not politically correct today! In addition to the insignias designed for NAS Jacksonville, other Disney insignias were designed for other station-based operations. The Gunners School at Yellow Water had their own insignia designed, showing a worried Donald Duck on a target, surrounded by bullet holes. An unidentified fighter training squadron at the station had an eagle on top of a Navy anchor, approved in March 1944. Naval Air Auxiliary Station Municipal #1, Operations, had an insignia approved in April 1945 showing a cat in a raft with his back legs spread, with a slingshot attached, pulled back getting ready to fire a bomber aircraft. Original Disney insignias made into patches if located today are quite rare and very valuable!

After the war ended, NAS Jacksonville was without an official insignia until the current triangle design was approved in February 1963.

FIRST STATION NEWSPAPER

Anyone asked what the official newspaper of NAS Jacksonville is would immediately respond the "*JAX*

First edition of the Jax Flyer, April 25, 1941.

First edition of the Jax Air News, April 1, 1943.

me dated October 20, 1994. "In November 1940, I was assigned as a Machinist mate First Class and arrived at NAS Jacksonville from Coco Solo, Panama Canal Zone. When I reported to the station nothing was ready. We had to build metal lockers in our barracks, install furniture in the officers quarters and we had no line ratings. So I had to impersonate one and take a motor launch out to lay anchored buoys to tie the P2Y flying boats to. As there was no station paper, I requested permission to print and distribute one of my own as an after hours business enterprise. Conditionally, permission was granted, but I had to submit all copy for screening by Naval Intelligence." He published at an old printing shop in Jacksonville and accepted advertising ads to pay for the paper. He would accept all ads except alcohol or tobacco ads.

Only one complete set of the papers exists today, furnished to me by Mr. Lavender. A review of the papers would find little information on actual events occurring at the station, probably censured out by the Naval Intelligence group. As an example, the December 12 edition, just 5 days after Pearl harbor, did not mention anything concerning that event! The paper went from 6 pages initially to 4, then expanded later on to 12 pages until publication stopped.

The demise of The *JAX FLYER* is somewhat a clouded issue today! At the reunion of early station aviators I attended in 1995, members told me the paper was folded due to the way advertising was being accepted and the Commanding Officer wanted more control of the paper. The paper shut down and the Public Affairs Officer was directed to start a new paper and control it on station. Thus, the *JAX AIR NEWS* was born. Mr. Lavender in his letter to me, however, had the real explanation of what happened. He stated: "In December 1944 I accepted the commission of Ensign so gave notice and stopped printing. I was told that these orders would be cancelled should I keep publishing but elected to go on down to the South Pacific and see what the war looked like." (Note: I assume he meant December 1942, as the *JAX AIR NEWS* started printing in May 1943.) In any event, the paper served a very beneficial purpose to the station personnel in the early years and Mr. Lavender and his small staff should have been commended for his excellent publication, one that would eventually lead to the *JAX AIR NEWS*.

AIR NEWS." It is true this paper has been the official newspaper for the station since Volume 1, Number 1 was issued on April 1, 1943. It has also been the newspaper of Naval Station Mayport; NAS Cecil Field; NAAS Lee Field in Green Cove Springs, and for numerous tenant commands at the station. But, in despite of its long history, it was not the first newspaper of the station.

On Friday, April 25, 1941 a new newspaper hit the station. Called *"The JAX FLYER"* it lasted in publication until the last edition was published on Friday, March 12, 1943. This weekly paper covered a small amount of base news, but mainly covered sports activities at the station and covered stories of general interest. Kermit Lavender actually started the first paper. He explained how it occurred in a letter to

STATION LICENSE PLATES

In 1941, NAS Jacksonville started issuing license plates to drivers who registered their cars with the station. There were only about 1,200 plates issued for 1941, and one of these plates in any condition is a rare find today. License plate issues continued every year after that except 1948. Plates were not issued for that year, but apparently the 1947 plate was accepted. In 1949, plates were again issued. In December 1952, windshield decals replaced the long familiar license plates, and windshield decals continue to be used today. The color code of the windshield decals today is as follows: Blue is for an officer; Red is for an enlisted; Green is for a civilian employee and White is for a contractor. Shown are some plates that have survived. It is not known what the colors or symbols mean in the way of officer or enlisted, if anything. The plates apparently were different colors every year, and on some years a Navy anchor was used with one color on the plate while the same year may show an aircraft propeller with another color of plate.

1941 License Plate
(From collection of Howard Hinz)

1945/1946 License Plate
(Author's collection)

1946/1947 License Plate (Anchor)
(From collection of Howard Hinz)

1946/1947 License Plate (Propeller)
(Author's collection)

1952 License Plate
(Author's Collection)

NR-1 Ryans and N2S Stearmans from Squadron VN-11 are stacked on their nose in preparation for an approaching hurricane. The Stearman N2S-3 #22 was the first aircraft ever reworked by the Overhaul and Repair Department (Naval Aviation Depot today.) (U.S. Navy)

A PBY, which made an emergency landing in the Trout River, was able to miss power lines and fly back to NAS Jacksonville in January 1944. A total curfew, requiring all military personnel to be off the streets of Jacksonville by 10:00 p.m., was also announced. However, it only lasted a few days.

The same Lt. Walsh, who had previously given lectures at NAS Jacksonville, was assigned to the station as an instructor in February. One of his first duties upon reporting to the station was to go to Washington to receive the Congressional Medal of Honor from President Roosevelt. Meanwhile, the five-cent movie charge, which had only been in effect for three months, was reduced to "free" in February at the station's three theatres.

In March, 150 aircrewmen received their wings during a coast-to-coast NBC broadcast with Bob Hope speaking at the ceremonies. Calling himself Bob "Jacksonville Naval Air Station" Hope, the screen star told the audience that "an aircrewman in combat with the enemy can crowd a lifetime of service to his country into 30 seconds." Also in attendance at the ceremony was Vice Admiral John S. McCain.

In April, Captain Walton W. Smith took the helm from Captain Gavin, who was assigned to sea command. Just a few days later, the entire station was saddened by the sudden death of Secretary of the Navy Frank Knox. Admiral Mason had another speaking engagement at NAS Jacksonville on May 27, 1944, and in one of his more humorous comments, stated "{we} lost 5,000 mince pies and thousands of doughnuts in the bakery which went down with the Hornet."

In June, all civilian employees of the station with more than one month of service were given a "Civilian Service Recognition Emblem" to wear. Developed due to numerous requests, the emblem was very popular and worn proudly by the station's 3,500 civilian employees. Captain Walton Smith, who became the fifth Commanding Officer on April 17, 1944, succeeded Captain Arthur Gavin. Brief ceremonies were held in the presence of department heads only.

The Ordnance and Gunnery Department celebrated the fact that over 250,000 students, firing over a million rounds with pistols, machine guns and shotguns, had done so without one accident, a truly remarkable record, also in April.

Jacksonville Municipal Number One was commissioned on July 1, 1944. The purpose of this station was to commence training in the PB4Y-1 Liberator Bomber. Municipal Number One at Jacksonville offered one of the longest runways on the East Coast, plus the advantages of Florida flying weather.

Mr. Alfred Sullivan visited my office on April 29, 1994, and told me the following story concerning life at boot camp on station during this time period. "I arrived by train at the Jacksonville terminal, then rode a bus to the station. A new class would arrive every week and there were 66 recruits per class. The school was located in the west area of the station {in the area where the Commissary Store/Navy Exchange complex is today}. When you checked in, you never left this area for the entire six weeks of boot camp. You learned basic Navy." He noted that for extra money, the Chiefs were running a scam where one could "rent" radios and clothes hangars that should have been provided for free! "The area of the boot camp was directly in line with the NE/SW runway and you could hear Corsairs and Hellcats flying about 100 feet over your barracks at night. It was most uncomfortable, combined with the August heat and humidity. Sometimes the platoon had to run at 1:00 in the morning. The running would consist of laps around a small area. Only on the last few days of the school did the recruits leave the compound to go to the parade grounds for the graduation ceremony, then immediately off the station on buses to their next assignment."

In October, an aerial searchlight course was established at the Naval Air Technical Training Center. The two-week course stressed the installation and maintenance of aircraft searchlights. The searchlights were being used by Navy torpedo and patrol planes, and were so powerful that it was possible to read a newspaper by their light from more than a mile away.

The worst crash for the Navy in their long 60-year history in Jacksonville occurred in October. Although the exact date is still not known, the following story is told by Thomas Garrett in a letter to the author dated February 5, 1994. "I did not see the midair collision myself, but watched briefly from the seaplane ramp while crash boats recovered some debris. As best as I can recall, it occurred near midday. Skies were only partly cloudy so weather was not a factor. I was told the collision was between a Vought F4U Corsair and a Douglas R4D (transport). I was told a number of the passengers were Navy nurses." (Note: The number of fatalities was 17, including the two pilots and 15 passengers.) This story came to light as I was reading an early official station historical report from the Public Affairs Office at the station written in November 1944. It stated the PAO was very successful in keeping this story out of the local press, even with the flaming debris covering the St John's River. The station would not see a crash as bad as this one again until 1983.

On October 26, over 1,700 civilians converged on the station to take refuge from a hurricane. Some 262 aircraft had already been evacuated from the station seeking refuge at Robin Field or Cochran Field in Macon, Georgia, or Daniels Field in Augusta, Georgia. The seaplanes went to Norfolk, Virginia. The path of the storm further chased the planes from their sites in Georgia to sites in Pensacola (Florida), Arkansas and even New York! All planes returned on 28 October except one Corsair, which crashed at sea and was never found. Total station damage amounted to $26,000. Supply Building 110 was converted into a disaster shelter where the Navy supplied its guests with beds, and even a kennel for pets. As the shortage of civilian workers grew acute, passes were established and passed out to all base employees. The blank passes were then filled out by the employee and could be given to prospective employees to get them easily into the base for interviews. Captain M. Cloukey, of the Assembly and Repair Department, desperately needed 1,800 employees to keep up with the war time workload.

In November, President Roosevelt was reelected, and Captain Herbert Regan relieved Captain Smith in brief ceremonies at the station. Captain Regan stated: "I know your reputation, I have seen your products and I have been favorably impressed." Also in November, chemical foam was first introduced to fight aircraft fires when the Airplane Crash Fire Rescue School was established by the Naval Air Technical Training Center.

An interesting study conducted at the station in 1944 found more than half the personnel smoking more than ever. The reasons for increased smoking were "Prospects of overseas duty," "Flying

Fatigue" and "Couldn't smoke at a civilian job."

The first all Marine squadron MF-OTU was established from the old squadron VF-4 to train Marines in fighter tactics. In December, a 60-bed institution as a unit of the Naval Hospital was opened for use by Navy dependents. The charge for an overnight stay was $1.75.

1945

1945 began with a visit from Brian Donlevy, a famous movie star of the day. A contract was also awarded to construct a Scrap and Salvage Building for $146,000. Today the structure is the center of the Defense Reutilization and Marketing Office, at U.S. 17 and Collins Road. Meanwhile, the first nine holes of the golf course were already in "full swing" and plans were being made to add nine more. How were the additional holes added? German prisoner of war labor!

Cary Grant visited the station in February much to the delight of his numerous fans. Squadron VPB-OUT #1 was also receiving their first delivery of the new seaplane PBY-5A's. The Crash Fire School was discontinued, after sending numerous trained rescuers throughout the Navy. March saw an enterprising enlisted man at NAAS Cecil Field start that station's own newspaper called "The Beam." It folded after 29 weeks of publication. Rear Admiral Davison replaced Rear Admiral McFall as Chief of Naval Air Operational Training in April. The growing shortage of housing got some relief with the construction of 140 additional units at Dewey Park.

The station observed victory in Europe with business as usual on May 7. The war in the Pacific against Japan had been what the station was really providing pilots for, and that continued to be their main focus. The war in Europe had been thought of as much more of an Army war.

The indoor swimming pool, designed to be used for aircrew instruction purposes, opened in June as the first German POWs were arriving. Captain Anthony Brady replaced Captain Regan and a new salvage furnace was constructed in July. The furnace melted 30,000 pounds a day into 25-pound ingots. Wrecked and salvaged aircraft were the main source of the melting operation. An additional $1.4 million worth of other materials were also salvaged from the same aircraft.

July 27 saw the establishment of a Flight Instructor School. A squadron of 83 aircraft was assigned for use by this school. The purpose of this school was to ensure a qualified supply of naval flight instructors to teach all aspects of fighter tactics and multi-engine operation.

Finally, peace came with the end of the war in the Pacific on August 15. All flying was immediately secured and all personnel were given the day off. Over 11,000 pilots and 10,000 aircrewmen and 30,000 gunners had received training at the station in support of the war effort. NAS Jacksonville was quickly designated as a personnel separation center, with Captain Gilchrist B. Stockton as its Commanding Officer. By the end of August, the air station gates were opened to the community for the first time. The public flocked in to see the displays of aircraft, men and materials that had helped to win the war. The entire base was opened up to displays and tours. An armada of aircraft flew over downtown Jacksonville. Admiral Marc Mitcher came to town and paid tribute to the great naval aviators.

Visitor regulations were reduced along with station working hours. The Naval Air Technical Training Center's fate was now up to Congress. At its peak in June 1944, 5,275 Navy students, 4,161 Marines and 55 foreign students had been in training.

By September, plans were being made for disposition of the Belmore, Bostwick, Branen, Carlisle, Fernandina, Foremost, Francis, Jacksonville Municipal No. 2, Palatka, Maxville, Middleburg, Mile Branch, Paxon, St. Augustine, Switzerland, Trout Creek and White House Practice fields and bombing ranges. Plans were also announced for all Waves to be separated by September 1, 1946. By October 1, 1,000 personnel had been returned to civilian life. On October 30, 531 officers and men were separated on one day and typists were urgently needed at the separation center. By December 6, as the point drop continued, over 25,000 personnel had been discharged.

The station's anniversary was formally held for the first time on October 15, 1945, on its fifth year. A parade was held in downtown with many floats from the station.

The Naval Air Operational Training Command underwent a name change to Naval Air Advanced Training Command (NAOTC)in November. This change was made by "The revision of the overall Naval Aviation Training program made necessary by adjustments to peacetime standards." PBY training was halted on November 13. The sight of seaplanes landing and taking off in the St John's River would soon be history as plans were being made to transfer the PBY's away from NAS Jacksonville.

December 5 saw the station helping in the search for the "Lost Patrol." These were the five TBM Avengers known as Flight 19 that left sunny NAS Fort Lauderdale, only to become disoriented in the Bermuda Triangle and never heard from again. Their last words were "It looks like we are entering white water...We're completely lost." As the search was coordinated from the NAOTC at NAS Jacksonville, a PBM Mariner flying boat was dispatched from NAS Banana River, only to also never be heard from again. For five days over 250,000 miles of ocean were searched. A Naval Board of Inquiry was formed and the final conclusion was "We are not able to even make a good guess as to what happened."

The year closed out with the disestablishment of NAAS Green Cove Springs on December 16.

THE GERMAN POWs ARRIVE

On Saturday, June 4, 1945, 500 German Prisoners of War (POW) arrived at NAS Jacksonville. They were quartered in a fenced-off, well guarded compound in an area occupied by the Recruit Training Center. The prisoners were transferred from the POW Camp at Aliceville, Alabama, and actual control of them was given to Captain George W. Gresham of Camp Blanding. Most of the prisoners had been captured in the previous two years and all were enlisted personnel. Building 411, which served for many years as headquarters for Readiness Command Region 8, was used as the mess hall for the prisoners of war. The new reserve center building at Birmingham and U.S. 17 is now located there. The POWs were housed in Buildings 463, 465, 466 and 467. By October 9, there were 1,645 POWs in the station compound, even though the capacity was listed as 1,000. Camp Blanding was originally designated as the only site in Florida to receive POWs, but as 1945 started and the Camp

German POWs arriving by train near the NAS Jacksonville Main Gate, June 1945. (National Archives #80-G-353582)

received more POWs than it could hold, branch camps were established and NAS Jacksonville became the main extension camp to Camp Blanding.

The POWs were available to all departments and tenants of the base and groups of ten or more could be "checked out" as long as one guard was provided for each group. One of the first tasks the POWs were assigned was to remove the spur line of the station railroad which ran along the north perimeter road to the sewage/water treatment area. Shortly thereafter, a large contingent was used to organize the base dump which contained mostly contractor's construction debris. The POWs segregated the dump materials, creating piles of 2X4 lumber and other materials that were in turn used by station maintenance personnel. Other POWs were used by Public Works Transportation to rebuild speedometers of cars and trucks—a task at which they excelled. The POWs also lined the drainage ditch that runs north and south through the middle of the base today, helped to construct the back nine holes of the Golf Course and did duty in the mess halls at the Naval Air Technical Training Center.

Mr. Foster, who supervised the POWs, said "It was hard not to feel sorry for these men. Most of them stated they surrendered during the African Campaign, some searching out American forces in order to surrender."

Since some of the POWs were aviators, they liked to watch the American aircraft flying at the base. The POWs were afforded more comforts than most American POWs. They had their own German Pastor, and music, movies, radios, sports and a library were made available to all. By May 1, 1946, the POWs were gone. They later actually had a reunion at the station, which many attended. This was because group of the POW's actually remained in the United States after the war and raised families here.

1946

As the Curtiss Seahawk was replacing the long familiar PBY Catalinas in 1946, NAS Jacksonville was among the stations to be named as a Ready Air Reserve Center. This was a plan developed by Admiral Mitcher to train approximately 28,700 Navy and Marine men. Cecil Field was being reduced and a Helldiver squadron (VSB) was transferred to NAS Jacksonville. The JAX AIR NEWS was cutting the size of the paper from 12 to 8 pages. The last PBY left the station on April 9, bound for NAS Pensacola. The Navy Department stated that plans had been made to start an aviation-training program by July 1, 1946, to maintain the combat-hardened pilots who were separated. By May, the newly formed

flight exhibition team now known as the "Blue Angels" was in practice. Also in May, the Seventh Naval District Headquarters was moved to NAS Jacksonville. May 27 saw another 45 aircraft damaged at NAAS Cecil Field as a freak hailstorm hit the station. Three aircraft were damaged beyond repair.

It had been announced that newly commissioned ensigns of the Naval Academy would be coming to NAS Jacksonville for instruction in aviation and flight procedure. The first group arrived June 10th just as a railroad strike was crippling commerce.

The Blue Angels flew their first show at the Craig Field dedication on June 15 and 16 and the Naval Air Reserve Program started on July 1 at Cecil Field in Hangar 14. A Marine Reserve unit, VMF-144 was added on 28 July.

The ferry service, so popular during the war, was down to two motor launches by September, and the Separation Center, after separating over 100,000 personnel, closed for good. General Wainwright visited the station.

The Reserve Training Unit moved from Cecil Field to NAS Jacksonville and held an open house in front of 6,000 visitors. Leaving the station in September was squadron VPB-208. This squadron arrived in May 1945 flying the big PBM Martin Mariners. They originally had 15 of the big flying boats, but were down to 9 when the squadron was transferred to Trinidad. With their departure NAS Jacksonville was left without any squadron giving training in multi-engine seaplanes.

October 15 saw no celebration of the station's sixth birthday. There was no half-day off like the year before, and no events to celebrate. The station was in the grips of reconversion to a peace-time status, strikes and shortages, and few people even realized it was the station's birthday. A power shutoff in Jacksonville crippled the station for a day in August. Navy Day was observed with a big open house for the last weekend in October, though. An even bigger event was the authorization of civilian clothes to military personnel while on liberty.

Herlong Field was given back to the city of Jacksonville on November 19, after years of training many Navy aviators. The station had originally acquired the 1,500 acres in August 1941 and had built the runways that still exist today.

THE "BLUE ANGELS" FORM AT

THE "BLUE ANGELS" FORM AT NAS JACKSONVILLE

edited by Butch Voris

World War 11 had ended. The military was down-sizing. The Navy's aircraft carriers had played a prime role in the Pacific, but were now back in their home ports far from public eye. The critical role of carrier operations was in danger of fading from public memory.

On April 24, 1946, Admiral Nimitz, Chief of Naval Operations, directed the establishment of a Navy Flight Exhibition Team to represent the Navy at military and public air shows. The team would form, train and base at NAS Jacksonville within the Naval Air Advanced Training Command. LCDR R. M. "Butch" Voris was selected as Officer-in-Charge and first team leader. He would select pilots, aircraft and support personnel from within the Training Command, and work out an air show flight routine. This was to be done on an urgent basis. Guidance was minimal. He was also directed to conduct training out of public sight. The Grumman F6F-5 "Hellcat" was selected over other fighters within the training command. It had been the primary carrier fighter during the war. It possessed honest flight characteristics, especially in the inverted slow speed regime, this being critical in close formation low altitude maneuvers. Four "Hellcats" were sent to the Assembly and Repair (A&R) department. Fifty pounds of sheet lead was sweated around each tail hook to bring the aircraft back into balance. The aircraft were repainted a lighter blue, with US NAVY and tail numbers in gold.

The four sparkling Flight Exhibition Team "Hellcats" rolled out of the A&R paint shop on May 9, 1946. Only recently (1997) did Butch receive a phone call and letter from the A&R painter asking him, "Who did you know to get all that gold leaf'? Butch was surprise to learn that actual gold leaf had been used for the gold lettering and numbering of the aircraft.

By this time Butch had selected pilots and support personnel. He had already developed a show routine which would be further modified and fine tuned as training progressed. It would be a grunt, turn and pull routine that would last 17 minutes and flown and positioned directly in front of spectators. He wanted hard impact. It was built on a belief that the routine had to be hard hitting, that is "get it up - get it on - get it down." It was destined to do just that.

The basic formation was built around three aircraft flying in vee and echelon dispositions. Rolls were done in both around the formation axis and with each aircraft around its own axis, the latter requiring the pilot to roll blind for one half of the roll. Although higher risk, it was indeed dramatic. The flight demonstration opened with a loop followed by a direct pull up into a Cuban Eight. This show opening started below 100 feet, with the loop topping out at 2,800 feet followed by varied low altitude echelon and inverted flight maneuvers. A crowd thrilling fighter to fighter dog fight was conducted half way through the demonstration.

The team initially consisted of Lieutenant Commander (LCDR) R.M. "Butch" Voris, USN, Officer in Charge and Flight Leader, LCDR L.G. Barnard, USNR, Lieutenant (LT) M.N. Wickendoll, USN, LT (JG) Melvin Cassidy, USNR, LT Alfred Taddeo, USN, and LT (JG) Ross Robinson. LT (JG) Billy May and LT (JG) Gale Stouse shortly joined the team.

The team was established in the Instructors Advanced Training Unit (IATU). Rear Admiral Ralph Davison, Chief, Naval Air Advanced Training exercised operational control, with Commander Dan Smith, Single Engine Training Officer, having day-to-day operational responsibility.

The Team was given a general training airspace over a desolate area west of NAS JAX. Butch relates that "if we had a mid-air crash, it would be out of view of the public, and that crash evidence would be destroyed by fire and alligators." Herlong Field as well as

The Hellcats of the future Blue Angels parked at Craig Field prior to their first air show performance, June 1946. (Mernard Norton)

The Blue Angels stand in front of their new F8F Bearcats at NAS Jacksonville in 1947. Flight Leader Butch Voris is in the middle. (U.S. Navy)

62

drainage ditches were used as training flight lines. "Unlike today, we had no restriction on flight altitudes. We were able to work as low as 50 feet."

"Our training schedule was intense. It was a matter of working-up step by step." Initially, Butch worked with each wing man separately, building trust, confidence and proficiency in close-in formation show maneuvers. Then he brought them all together as the next step, still working at safe (ball out) altitudes. This was followed by bringing them on down gradually to show altitudes. Between formation flights pilots practiced individually further increasing their confidence in performing at near ground level.

In early June, Voris advised Commander Smith that he believed that the Team was ready for their first evaluation. Herlong field would be the site, with Commander Smith flying an F6F out to Herlong, landing and establishing radio contact with the Team. His only words were, "let's see what you've got". With that, Butch rolled his Team into the flight line and the show began. Upon completion, Commander Smith's cryptic words were, "I'll see you guys back at JAX".

Voris believed that they had flown a good routine. Still, he didn't know just what was expected of them. Always being in the cockpit, he had never seen the Team perform from the ground. It was before the era of video cameras, etc. On landing, we were greeted with the comment, "you crazy sons of bitches are going to kill yourselves, but I like what you've got. I'll recommend that the Admiral see your routine tomorrow afternoon". The performance for Admiral Davison went exceptionally well. He arranged for Vice Admiral Frank Wagner, Chief on Naval Air Training to observe our flying at NAS Pensacola. This demonstration went well until the dog fight which concluded in the dumping out of a sawdust/sand filled dummy enemy pilot with parachute. The static parachute line failed and the dummy and unopened chute impacted much too close to the Admiral and senior staff who were positioned in chairs out on the aircraft parking ramp. Voris expected that they would be ordered to immediately discontinue the demonstration and land. In the remarkable absence of such an order, he continued on and completed the show. On landing, and fully expecting Admiral Wagner's wrath, he instead listened to the admiral's first words, "I suggest that you move the shoot down show took off, it was realized that the Team needed a popular name. The station Newspaper, the JAX AIR NEWS started a contest to name the new team. The paper announced the contest on June 27, barely two weeks after the initial performance. Names submitted should be short and descriptive. Names poured in. All fell short of what the team though it should be. Although the name Navy Blue Lancers was "brought to Butch's attention" by senior officer whose son had submitted the name, he was looking at a name that would compare with the Sea Hawks and the Three Musketeers of early aviation fame.

A night or two later, the team was sitting around Voris' Senior BOQ room discussing a coming show in New York, after which the Team would turn in their "Hellcats" for Grumman F8F "Bearcats".. All of a sudden, "Wick" Wickendoll excitedly exclaimed, "I've got it, I've got it ...... Wick", while leafing through the June issue of the New Yorker Magazine, came across the listing of the Blue Angel Dinner Club in downtown Manhattan. It was a natural - The Blue Angels. Now, the dilemma, how would the Team get this name to stick? The Team with the help of the aviation press at the "World' Fair" aviation meet in Omaha, Nebraska on July 19-2 1, broke the name in post show press releases. Voris remembers the

The Blue Angel team at NAS Jacksonville, June 17, 1947. A Bearcat is the team's plane. (U.S. Navy)

"Beetle Bomb" used by the early Blue Angels. (Don Brammer)

chilly reception on the Team's return to JAX.

A NAS JAX Open House was set up to celebrate the newly formed Naval Air Reserve's move from Cecil Field to NAS JAX. During the air show the first tragedy struck the team. As the JAX AIR NEWS described the accident, "Horrified, the crowd stood transfixed as the pilot, completing a low altitude pull out from a one half Cuban Eight, experienced a wing outer panel failure causing him to roll into the mat directly in front of the spectators. Death for Lieutenant R.F. Robertson was instantaneous.

The Team continued to be based at NAS Jacksonville until their transfer to NAS Corpus Christi in October 1948: Only NAS Jacksonville, NAS Pensacola and NAS Corpus Christi can claim a Blue Angels Flight Demonstration in every type of aircraft that the Blue Angels have flown throughout their history!

1947

On January 10, a strange sight appeared at the station. A Phantom jet arrived for a two-day stay. The jet, piloted by Lieutenant Commander W.W. Kelly, made a stopover on its way to the Miami Air Races. The flight was made the next day in 42 minutes, averaging 420 miles per hour. Early in 1947, a three-day Regional Safety Conference was held on the station and the F4U Corsair became the main aircraft to be overhauled at the Assembly and Repair Department. The A&R Department was also opening a new indoor firing range to test overhauled machine guns. It was the first of its kind in the U.S.

The Blue Angels would return to Craig field for another air show on April 26 and 27. The team had traded in their Hellcats for faster and more maneuverable Bearcats.

Captain Brady was relieved on June 1 by the Executive Officer until Captain Herbert Duckworth became the Commanding Officer some two months later. Captain Brady left for duty in Adak, Alaska, and one short month later, the popular Commanding Officer became the first past CO to die when he was struck by a heart attack. As naval stations were closed all over, the personnel at NAS Jacksonville became concerned they might be cut until a CNO Directive in June stated that NAS Jacksonville would be maintained in a full operational status. As stated in the directive: "The primary mission of subject station will be to provide facilities and support for the Advanced Flight Training of Naval aviators as directed … and for a Class "A" Aircraft Assembly and Repair establishment." As a secondary mission, the station was to provide support and facilities for Headquarters, Commandant Seventh Naval District; Headquarters, Chief of Naval Air Advanced Training; Naval Air Technical Training Center; Naval Air Reserve Training Unit; Naval Air Transport Service Detachment, and assigned fleet units.

In July, NAS Banana River was losing squadrons as two of its three squadrons were reassigned to NAS Jacksonville. Coming to the station were a torpedo bomber squadron flying Avengers and a squadron flying Martin Mariners. NAAS Cecil Field received a squadron of F6F Hellcats. NAS Banana River was to be closed by August 1. Cecil Field's new squadron would not have a home for long, though, as the auxiliary station was to be closed by October. As a part of the city's festivities, naval reserve aircraft flew over the Gator Bowl and formed the letter 'F' (for Florida Gators) and 'G' (for Georgia Bulldogs) at the annual Florida-Georgia football game. Air, rail and bus ticket offices were also opened during August for the benefit of station personnel for the first time. The Naval Air Gunnery Range at Yellow Water was closed effective October 6 and a Hellcat crashed just south of Lee Field in Green Cove Springs on October 30. The pilot was practicing night landing maneuvers by covering his canopy and using instruments. As he looked out of the corner of his eye, he saw the tops of pine trees going by. It was too late to act and the fighter went into the swamp. The pilot received minor injuries and lay waiting for the ambulances from NAS Jacksonville. The guns and some radar units were removed from the plane, along with the injured pilot. The plane remains in the swamp today, somewhat deteriorated.

Finally, the Navy Wives Club was formed and had their first meeting on December 3.

1948

In January, the Naval Air Technical Training Center announced its move to Memphis and by February 13, instruction of personnel stopped. Only two small schools and the operational trainer remained. The center of the station, once a bustling training activity, resembled a ghost town by March.

By April, the station employed 3,645 civilian employees and an additional 900 were added in May. Captain Duckworth was relieved by Captain Malstrom in June and the first jets, Phantom IIs, landed on the 13th, drawing big crowds. The next day two PBY's returned to the station for Search and Rescue (SAR) duties and the PBM's that arrived last year were leaving the station for assignment to Corpus Christi. The SC-1 Curtis Seahawk would be the only seaplane left which was used for training.

May 27 saw a ceremony at the station to award 16 pilots their wings. This was the first group to get their wings here since 1943.

A major reorganization of station departments was completed in June. The Yard Department was completely reorganized and the Fire Department was transferred to Public Works, where civilians replaced the former all-military functions of Fire and Rescue. Air Operations was also designated to become a department of NAS Jacksonville, rather than remaining under the soon to be disestablished Aviation Training Division. Approval was also given by the Chief of Naval Personnel to establish a Chief Petty Officers' Mess at the station.

July saw a name change for the Assembly and Repair Department, as it became the Overhaul and Repair Department. At this time, Corsairs were the major aircraft being overhauled at the department.

By August, it was announced in Washington that NAS Jacksonville, along with the auxiliary stations at Cecil Field and Mayport, would be the home of carrier groups. This would make up for loss of the Seventh Naval District, which was absorbed by the Sixth Naval District in Charleston, and for the loss of the Naval Air Advanced Training Command. Also scheduled for departure were the Blue Angels.

On October 21, Midshipman Jessie L. Brown became the first black in the Navy to receive the wings of a naval aviator in a ceremony at the station. The station Commanding Officer attached the golden wings to his chest; a photographer was on hand to record the event and the public information officer released the story and picture the next day: First Negro Naval Aviator. Strangely missing was any coverage of the historic event from the station newspaper the JAX AIR NEWS. Brown would go on to fly his plane in Korea and be shot down. Another aviator would crash his plane by Brown to try and rescue him. A rescue helicopter came and they could not extract him from his plane. Bleeding badly, he eventually bled to death. After the rescue helicopter departed, another Corsair bombed the crashsite to keep the plane and the body from falling into Korean hands. When NAS Jacksonville dedicated Patriots' Park in April 1996, one of the three Medal of Honor recipients who came to the ceremony, Captain Hudner, was the pilot who tried to save Jesse Brown. Brown was posthumously awarded the Distinguished Flying Cross, and the Purple Heart. He was further recognized when the keel was laid on April 9, 1971, for the Jesse L. Brown (DE-1089), a new destroyer escort ship which was named after him.

Cecil Field was reopened as an auxiliary station by November 1, along with Opa Locka Air Base, Miami, and NAS Fort Lauderdale. Secretary of Defense James Forrestal, who had always supported the establishment and continuous operation of bases in Jacksonville, made the announcement. In addition, NAS Jacksonville emerged as a major fleet air base with the announcement Carrier Air Groups were to be assigned to the station. November

1 also saw the Commanding Officer of NAS Jacksonville take on additional duties as Commander of the Sixth Naval District. Commander Shea become the Commanding Officer of NAAS Cecil Field and the Naval Air Advanced Training Command transferred to NAS Corpus Christi. Prior to the Blue Angels leaving for Corpus Christi they flew one last show at the base where they were formed on November 8. On November 10, the Carrier Air Groups (CAG) started arriving. The first to come was CAG-8 with its assigned squadrons of VF-81; VF-82; VF-83; VA-84 and VA-85. Fleet Aircraft Service Squadron SIX (FASRon 6), also arrived at the station.

1949

CAGs were starting to arrive in force as 1949 started. CAG 17 started arriving on January 14, from Quonset Point, Rhode Island, assigned to NAAS Cecil Field. The attached squadrons were VF-171, VF-172, VF-173, VA-174 and VA-175. Arriving on January 15 at NAAS Cecil Field were the Phantom jets of VF-171, the first Navy east coast jet squadron. CAG 4 arrived from Norfolk to NAS Jacksonville on February 15 with squadrons VF-41, VF-42, VF-43, VA-44 and VA-45. On February 24 CAG 1 arrived at NAAS Cecil Field with squadrons VF-11, VF-

12, VF-13, VA-14 and VA-15. The CAG had 125 officers and 500 enlisted assigned. NAS Jacksonville's CAG 8 and their assigned squadrons were in Washington on February 20 flying for the inauguration of President Truman. The squadrons, flying F4U Corsairs and AM-1 Maulers, were part of one of the most massive air armadas in the nation's history.

A Philippine Mar's aircraft loaded with equipment from Carrier Air Group 13 arrived in April and squadrons VF-131, VF-132, VF-133, VA-134 and VA-135 arrived at NAS Jacksonville shortly thereafter. CAG 13 was the fifth and final CAG to be assigned to the area. With the arrival of this CAG, Fleet Air Jacksonville now had 630 officers and 3,000 enlisted assigned between NAS Jacksonville and NAAS Cecil Field. On May 3, USS Midway was off the coast for four days so carrier landing practice could take place with the squadrons. The station again hosted a huge air show in May to commemorate the 30th anniversary of the first transatlantic flight by Navy plane NC-1. In the meantime, City of Jacksonville leaders were pushing for the development of a carrier basin at Mayport. (Of note today, some 50 years later, is the fact City leaders are once again pushing the development of the carrier basin, this time for Nuclear carriers.)

The Naval Air Reserve Training

The first jets, F2H Phantoms arrive at the station in June 1948. (U.S. Navy)

Units had also grown considerably, with 10 aviation units now in operation. The station received the 1948 Secretary of the Navy Safety Award in August, the first of many to come.

A hurricane hit in late August, but did little damage. The station's planes were evacuated as far away as Dallas. All safely returned. September 26 saw the huge R-60 Constitution land at the station, only one of two of these aircraft that existed. The Navy had little real mission for the big planes so the stops at the station were mostly for publicity. Squadron VA-135 became the last squadron on the East Coast to replace their TBM-3E's Avengers with new AD Skyraiders on September 20.

The Overhaul and Repair shops were also getting into the jet business in November as the first FH-1 Phantom arrived for overhaul. Fleet Air Jacksonville observed their first birthday on November 1. In one year the command had grown from a staff of three officers and a total of six squadrons (Carrier Air Group Eight and Fleet Air Service Squadron Six) to a present number of 34 squadrons and a personnel strength of 5,000. Included were the only two jet squadrons on the east coast. This command contributed $8 million to the Jacksonville economy. But just as quickly as the CAG's came, an announcement was made in late November that budget cuts would cause the loss of Carrier Air Groups 8 and 13, which were to be decommissioned by February 1, 1950.

As the 1940s were ending, times became more austere. Civilian employees were being cut and Fleet Air Jacksonville was about to lose Carrier Air Groups 8 and 13. On a somewhat brighter note, Fleet Air Wing Eleven, two VP squadrons and Fleet Aircraft Service Squadron 109 were scheduled to come to NAS Jacksonville, partly to offset that loss, with the forward detachments arriving on 24 November. Squadrons VP-3 and VP-5 were being transferred from San Juan, Puerto Rico, and Coco Solo, Panama Canal Zone. The seven planes of VP-5 (P2V-3 Neptunes) actually started arriving on December 9. Even with future squadron reductions right around the corner, the decade was ending on a positive note for the long-term future of the station. NAS Jacksonville had arisen from the woods and marshes to become one of the first line naval air stations of the world. This phenomenal growth would fortunately continue through the 1950s as the station strengthened its role in supporting naval aviation.

THE 40s IN PHOTOS

Admiral Marc A. Mitscher (right) was the special guest of honor at NAS Jacksonville's Fifth Anniversary, October 1945. (U.S. Navy)

Squadron VN-11 receives instruction in the NR-1 Ryan in 1942. (U.S. Navy)

Naval Hospital Jacksonville as seen in February 1942. (U.S. Navy)

President Roosevelt in front of the Administration Building while on a station visit, March 1941. (U.S.Navy)

Main entrance of the Assembly and Repair Department, December 1942. (U.S. Navy)

This PBY, assigned to VPB2#1, was set-up as an outdoor classroom for ordance training on August 6, 1945. The plane was nicknamed "Gravel Gertie." (U.S. Navy)

Left: Abandon ship drill is practiced on August 24, 1944. Notice student using the person's face as a step on the right. (U.S. Navy)

Bob Hope receives the honorary "wings" of a naval aviator while at the station in March 1944. (U.S. Navy)

Annapolis graduates take their turn on the station's training guns, June 29, 1944. (U.S. Navy)

Aircraft on the apron near the NAS Jacksonville Control Tower in 1944. Plane to the right is an R4D and a PV-1 is on the left. (U.S. Navy)

The Physical Training Department teaches sailors how to use their trousers as a flotation device, March 21, 1944. (U.S. Navy)

Right: A rare photo showing WAVES washing an SNJ assigned to the Naval Air Operational Training Command, 1944.

A 1945 aerial view of the seawall showing the PBYs on the ramp. (U.S. Navy via J.T. McCulley)

A PBY is parked on the ramp with engines running to simulate wind conditions for a movie on parachute safety in 1944. (U.S. Navy)

WAVES being taught CPR at the station pool, September 29, 1944. (U.S. Navy)

WAVES Aviation Metalsmith working on an SNJ at the Assembly and Repair Department, 1944. (National Archives #80-G-88213)

"Paddles" bring a plane in during WWII. A school to teach this technique was conducted at the station for many years. (U.S. Navy)

Yellow Water Gunnery Range, 1944. (National Archives #80-G-241916)

Left: The first helicopter visit to the station on November 28, 1944. (U.S. Navy)

Another engine is removed from an aircraft to be scrapped, July 1944.

Upon researching this photo, I discovered the machanic to the left was my grandfather, J. S. Robinson. He retired from the Naval Air Rework Facility in 1967. This shows engine removal, August 23, 1945, on an F4U Corsair. (U.S. Navy)

Bombsite training conducted at the station in 1944 trained hundreds of personnel. (U.S. Navy)

The fabric wing shop at the Assembly and Repair Department, December, 1941. (U.S. Navy)

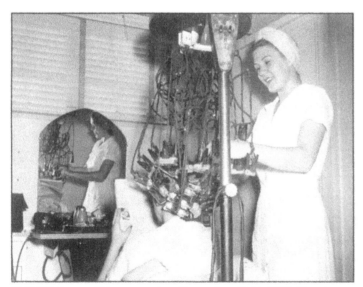

WAVES Beauty Shop at station, November 1944. (U.S. Navy)

This building, which was located at the corner of Allegheny and Albemarle Avenue, was used as the Marine's first barracks and eventually the first Chief's Club. (Florida Times-Union)

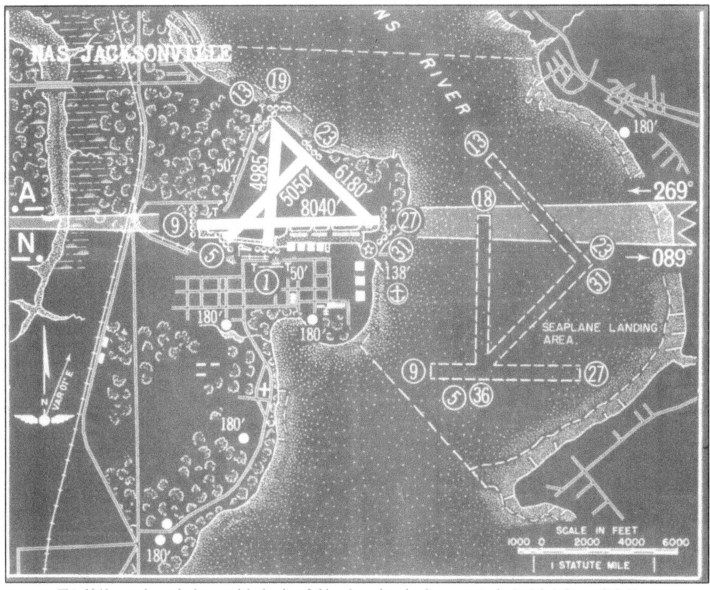

This 1941 map shows the layout of the landing field and seaplane landing areas in the St. John's River. (U.S. Navy)

ANNIVERSARY
PROGRAM
October 15th, 1945

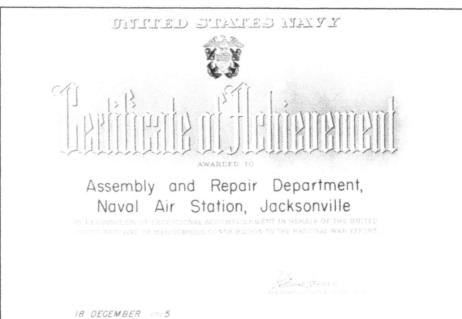

UNITED STATES NAVY

Certificate of Achievement

AWARDED TO

Assembly and Repair Department,
Naval Air Station, Jacksonville

IN RECOGNITION OF EXCEPTIONAL ACCOMPLISHMENT IN BEHALF OF THE UNITED
STATES NAVY AND OF MERITORIOUS CONTRIBUTION TO THE NATIONAL WAR EFFORT

18 DECEMBER 1945

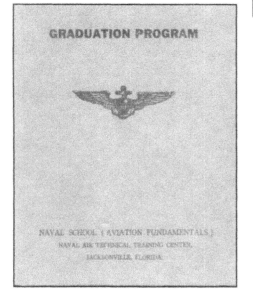

GRADUATION PROGRAM

NAVAL SCHOOL (AVIATION FUNDAMENTALS)
NAVAL AIR TECHNICAL TRAINING CENTER,
JACKSONVILLE, FLORIDA.

N. A. O. T. C.
JACKSONVILLE, FLORIDA

Certificate of Membership

No.

WAR BONDS

$1,000.00 CLUB.

This is to Certify that

holds a $1,000 share of stock in The Allied Victory
Corporation, and is hereby granted full authority to
smash all Axis persons conspiring against said
Corporation, and is furthermore Bestowed the title
of being a full fledged and honored member
of the $1,000 War Bond Club.

Date

WORLD WAR II AIRCRAFT DISPOSAL

The photos on this page from NAS Jacksonville show how excess aircraft were disposed of during 1944. Most were put on the scrap market. Some went for parts, but many followed the sequence shown here.

Below: During 1944, these FM Grumman Wildcats parked at the station were virtually worthless, as the market for scrap was gutted.

Bottom: Here the engine is removed, along with instruments.

Right top: Next, after a thorough steam cleaning to purge out all fuels, the wings were cut off.

Right center: The fuselage is then picked up by the crane and . . .

Right bottom: Loaded onto a barge at the station pier for the trip to be dumped at sea.

POST CARDS FROM THE 1940s

OS2U-2s IN FORMATION
N.A.S. JACKSONVILLE, FLORIDA

GENERAL ASSEMBLY
AERIAL VIEW LOOKING SOUTHWEST
N.A.S. JACKSONVILLE FLORIDA

ASSEMBLY AND REPAIR SHOP
N.A.S. JACKSONVILLE FLORIDA

ENLISTED MENS MESS HALL AND BARRACKS
N.A.S. JACKSONVILLE FLORIDA

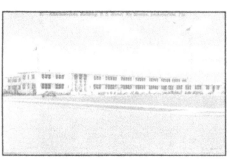

SNUG HARBOR

Kingfisher Squadron   *PBY Squadron*

Officers and crew of SNBs stand inspection at NAS Jacksonville. These aircraft were assigned to Squadron VA-ATU #4, March 1943. (National Archives #80-G-407296)

Above: A Curtis SC-1 Seagull is hoisted ashore on April 18, 1947. (U.S. Navy)

Right: An OS2U Kingfisher lands on the St. Johns River, March 1943. (National Archives 80-G-407866)

J2F-3 Grumman Duck assigned to NAS Jacksonville, January 8, 1942. (U.S. Navy)

This N3N-3 was the first plane to land on the still unfinished station runway on September 7, 1940. (Florida Times-Union)

Stearman, N2S-5, assigned to station, June 1945. (U.S. Navy)

NR-1 Ryan trainer. (U.S. Navy)

An OS2U Kingfisher heads down the seaplane ramp and into the St. John's River. (National Archives #80-G-407871)

This GH-1 Nightingale was assigned to station January 1943. (U.S. Navy)

PBY-2 assigned to the station for seaplane training on January 8, 1942. (U.S. Navy)

PBY-5A in landing configuration, assigned to VPB2, OTU #2. (National Archives #80-G-405451)

JRB assigned to station on January 8, 1942. (U.S. Navy)

JRF-5 assigned to station, painted white and sky blue, on January 8, 1942. (U.S. Navy)

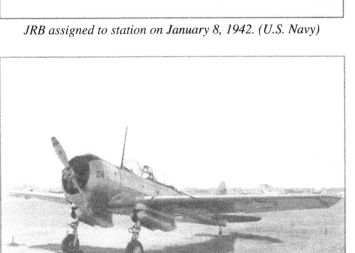

SNC-1 on ramp, January 1942. (U.S. Navy)

SBD-5 weather plane assigned to station on April 18, 1945. (U.S. Navy)

Beech SNB-2H assigned as the hospital plane. Photo April 18, 1945. (U.S. Navy)

The first helicopter to land at the station, a Sikorsky HNS, is shown here landing on November 28, 1944. (U.S. Navy)

An F4U Corsair catches the wire during landing practice, 1944. (National Archives #80-G-312367.

The first seaplane at NAS Jacksonville, this Consolidated P2Y, 1941. (Florida Times Union)

In one of the most colorful ceremonies ever held at the station, six Support Landing Ship (Large) Mark III (LSSLs) were transferred by the US government to the Italian Navy in July 1951. (National Archives #80-G-439753)

Officers and men of the Naval Air Reserve Training Unit salute the city of Jacksonville in January 1954. (U.S. Navy)

1950

The decade of the 1950s began in a most unusual way for many of the personnel stationed at NAS Jacksonville. Because Captain Alvin Malstrom, the Commanding Officer at the time, had determined that Casa Linda Lake needed to be cleared of the water-choking vegetation, "Operation Bulrushes" was initiated. Approximately 1,000 male sailors, divided into groups of 10 enlisted and one officer, were ordered to descend upon the previously neglected lake. Since it was a cold and rainy Saturday morning in January when the operation took place, female personnel remained on shore, serving hot coffee and doughnuts. The lake was completely cleared in only about three hours but the line of sick sailors was way out the door at the Dispensary on Monday morning! Casa Linda Lake was eventually restocked with game fish and has become one of the most popular recreation sites at NAS Jacksonville.

Meanwhile, Fleet Air Wing Eleven was completing the move to its new home station. Transferring with the Wing were squadrons VP-3 (from Coco Solo) and VP-5 and FASRON 109 (both from Roosevelt Roads). In addition, VP-23 was assigned to NAS Jacksonville from time to time, although the "Hurricane Squadron" (as VP-23 was known), maintained its primary home in Miami. Carrier Air Group Four and its associated squadrons left on January 4, for a cruise on the USS Midway to the Mediterranean. Other aviation activities occurring in February included the announced disestablishment of Carrier Air Group (CAG) Eight, and CAG Thirteen. Along with the CAG's disestablishment were squadrons VF-81, VF-83, VA-174 and VF 131.

NAS Jacksonville was growing. At the same time that base population was increasing, the Naval Hospital was facing severe bed cuts, going from 500 to 325 to 180 by July 4. The Veterans, however, through intense pressure, managed to increase their bed capacity from 20 to 40.

The Blue Angels, now flying the F9F "Panther," flew a show at the station on April 28 and 29. By June 30, however, the team was out of the air show business and flying combat duty in Korea as an Operational Squadron. NAS Jacksonville would not see them again for over two years.

Despite the increase in squadrons assigned to NAS Jacksonville, the Overhaul and Repair Department had to cut 100 employees from the payroll, along with six other air stations.

On the plus side, Congress finally passed a bill authorizing the carrier basin in Mayport in May. Also in May, for the first time in two years, student training started again as the "Advanced Aviation Electrician Training School" commenced at the station. On May 23, units from NAS Jacksonville participated in the Armed Forces Day Parade in downtown Jacksonville.

June 5th saw the first helicopter assigned to the station along with the first attempted rescue. A Sikorsky H03S-1 from HU-2 (based at Lakehurst, NJ) arrived for search and rescue (SAR) duty. Their first rescue ended on a bizarre note as a Jacksonville man thrown from his boat when it capsized in the St. John's river, refused to go into the strange new sight! The next rescue went smoothly, though, as they rescued Ensign John Neal of VF-74. He had crashed his F8F Bearcat at Lee Field in Green Cove Springs.

On 15 June, Captain Douglas Turner Day became the new Commanding Officer, relieving Captain Malstrom. He also assumed duties as Commander, Naval Air Bases, Sixth Naval District. The station's fighter squadrons were also changing hands as VF-11 and VF-14 also had change of command ceremonies. VF-14 was the Navy's oldest fighter squadron during that time.

It remained a time for cuts, as CAG Four was disestablished on 8 June. To make up for this loss CAG One and CAG Seventeen were transferred from NAAS Cecil Field to the station.

June 22 showed many the force of lightning. A lightning bolt struck Mason Field during the eighth inning of a baseball game. After the bolt, eight players and an umpire lay unconscious on the field. None suffered permanent injuries, however. On the same day, the Navy's newest bomber, the AJ-1 landed at the station.

In August, Cecil Field was ordered reactivated, scarcely a month after it had completed a previously ordered inactivation. As a fleet operational base, it reopened under Commander, Naval Air Bases, Sixth Naval District. By August 17, it was announced that $3.75 million would be spent to establish Cecil Field as a master jet base. Meanwhile, jet noise was becoming an issue with Jacksonville residents and the Navy defended its position by declaring "When flying over the Southside, pilots will remain under 700 feet to avoid commercial air line routes and possible collision." Unfortunately, August was not a good month for aviation safety at the station. Two ADs collided on August 4, a TBM crash landed on August 14 and an F2H-2 and an F4U crashed on August 15.

In September, the recently reduced Overhaul and Repair Department was busy with the F2H Banshee and TO-2 Shooting Star jets, the F4U Corsair and the R4D Transport. They set an impressive record for the quarter ending September 1950 by completing 281 overhauls. Construction of a Jet Maintenance Shop (Building 200) started in September. At the same time, the possibility that Navy helicopters would be overhauled at the Overhaul and Repair Department was being discussed. This would eventually occur the next year. Squadrons VF-44, flying F4U Corsairs, and VA-45, flying AD-2's, were established in September. VA-106, VF-44 and VA-45 then left NAS Jacksonville to go to Cecil Field as CAG Four was reestablished.

In October, the Overhaul and Repair Department again set an NAS Jacksonville repair record when 16 aircraft were completed and "sold" (accepted), as ready in one day. This one-day record would never again be equaled.

Squadron 861, formerly at NARTU, joined Fleet Air Wing Eleven in November. No other major events were recorded for the remainder of 1950.

1951

The year 1951 began with good news when it was announced that the Naval Air Technical Training Center would be reactivated in January. Nine schools were to be re-established at NAS Jacksonville using seventy station buildings. Plans were being made to train up to 12,000 students. Also in January the Navy was purchasing an additional 700 acres at Mayport for expansion of aviation facilities. The extra acreage will allow for expanding the runways to 8,000 feet from their current 4,200 feet. A young movie star visited the station and the Naval Hospital on January 20. He would be later known as President Ronald Reagan!

The 5th of February saw one of the largest planes in the world land at NAS Jacksonville. A KR-60 Constitution landed carrying cargo.

Thursday, February 15, Ted Mack and the Original Amateur Hour originated from the station. Also in February, it was announced that active duty would begin for the reserve squadron of VF-742. Shortly afterward, VF-741 and marine outfit VMF-144 also joined the ranks of active duty as the Korean Conflict heated up. Pilots from NAS Jacksonville based VF-12 also became the first pilots to qualify for night carrier operations when they landed on the USS Oriskany.

Patrol Squadron 10 (VP-10), was established as a new squadron on March 19 under Fleet Air Wing Eleven, along with the reestablishment of CAG Eight.

May 1st saw the commissioning of Naval Auxiliary Air Station at Sanford, Florida. Captain Day read the commissioning orders. Future squadrons would be relocated from NAS Jacksonville.

By July, entertainment was on the minds of NAS Jacksonville personnel as Morton Downey and the Three Stooges were performing at the Technical Center Auditorium (King Hall). The show was well received. Six mothballed vessels formerly located at Green Cove Springs were turned over to the Italian government in a colorful ceremony at the pier at NAS Jacksonville on July 25.

The possibility that helicopters would be overhauled at the Overhaul and Repair Department was being discussed in August. An appropriations bill had $9.9 million set aside for construction of overhaul facilities.

Aircraft squadrons continued transfers in September. Squadron VA-85 transferred from NAS JAX to NAS Quanset Point; VF-916 transferred to NAS Oceana; and Fleet Aircraft Service Squadron (FASRon) 51 was reassigned to NAS JAX.

1952

Early in 1952, two more squadrons, Composite Squadron Five (VC-5), flying the AJ-1 bomber and the P2V Neptune, and Fleet Aircraft Service Squadron Fifty-one (FASRON 51), a support unit, were assigned to duty at NAS Jacksonville. These units would form the nucleus of Heavy Attack Wing One. VP-3 received the first P2V-5 Neptune in the Atlantic Fleet in January. In February VP-10 was transferred to NAS Brunswick where they remain today.

On March 10, NAS Jacksonville became the center of hurricane hunting operations when VJ-2 was commissioned. It immediately took over the duties of squadron VP-23, based at Miami. The squadron was formed flying the P4Y-2 "Privateer", which arrived March 11. On April 10, Photographic Squadron VJ-62 was commissioned.

An innovative idea that was implemented in May was the Technical Training Center Telephone Lounge. A sailor could go to the lounge, give an operator the number he wanted, wait for the call to be placed, and then be directed to a phone to talk. When his call was completed, he would pay the lounge operator. The Center became very popular with base personnel. In June, the JAX Navy Federal Credit Union opened its doors at NAS Jacksonville.

On July 1, Cecil Field finally severed its ties with NAS Jacksonville to become a separate Naval Air Station. The changeover was done without fanfare, when Commander William Stevens, Commanding Officer, read the Chief of Naval Operations Directive in a simple ceremony. Based at the station were Carrier Air Group 10 (being formed), FASRON 9 and Composite Photographic Squadron VC-62. An FJ "Fury" jet arrived at the Naval Air Technical Training Center (NATTCen) also in July so students could start classes on jets.

In August, Captain Burnham C. McCaffree relieved Captain Day as Commanding Officer at NAS Jacksonville and work on repair of helicopters commenced at the Overhaul and Repair Department. Full-scale production was still months away, but minor repairs had already started back in March. Eventually, the HUPS (Utilities), HOS (Observation), and the rescue craft, HRS-1 and 2, would be repaired.

By October, the new jet "ejection trainer" was being used extensively. The trainer, one of the first ever built, eventually tested 20 pilots a day.

In late October, the first carrier to dock at Mayport was having problems. High winds and strong river currents caused cancellation of the U.S.S. Tarawa's arrival for three days. On October 30, 1952, the ship finally docked.

The air station won its second Secretary of the Navy Safety Award in November. VJ-2 received its first P2V-3W "Neptune" on November 10. The station finished the year with an air show featuring the Blue Angels, flying F9F-5's, on December 7.

1953

In February 1953, VP-741 was redesignated VP-16 at impressive ceremonies in Hangar 140. By June 1, the Navy Credit Union now boasted $40,452 in assets after its first year of operation.

In September, an F7U "Cutlass" arrived at the station for NATTCen training duty. In November 1953, a guided missile school started classes at the Technical Training Center, the first of its type in the Navy.

On December 15, 1953, VJ-2 (Hurricane Hunters) were redesignated Airborne Early Warning Squadron Four (VW-4). The Blue Angels also performed at Mayport for the first time on December 4.

1954

January started out with the naming of the first building at the station. The station auditorium was named "King Hall" after Fleet Admiral Ernest J. King. The building remained King Hall until it was demolished in March 1992.

In February 1954, VF-44 became the last squadron to switch from the F4U Corsair to the F2H-2 Banshee jet. At the same time, the Overhaul and Repair Department completed its final F4U overhaul. Since 1943, when the department first began overhauling the F4U, 2,872 had been completed over the ten-year period. For the thousands of Overhaul and Repair employees who had received, stripped, cleaned, disassembled, overhauled, repaired and put the F4U back together again, it was indeed a sad day. Nonetheless, the airfield was still a busy place with the Flight Control Tower handling almost 800 aircraft per day.

On April 30, two VP-3 P2V-5 Neptunes made the first successful Iceland to Jacksonville non-stop flight. Earlier in the year, a VP-3 flight had ended in disaster when the plane crashed in Iceland killing all on board. A landing signal officers school, established at Whitehouse under control of Fleet Air Wing Eleven, became another "first" when it opened in June 1954.

In December, VW-4 (formerly VJ-2), was preparing to shift from the P2V-5 Neptune to the NC121 Super Constellation. These planes would provide the greater speed and distance needed by the squadron for their hurricane hunting operations. Also, VP-16 became the last Atlantic Fleet squadron to trade in their P2V-2's for new P2V-5's.

1955

The Overhaul and Repair Department started 1955 with a bang by setting another record with its completion of the first S2F Tracker Antisubmarine Warfare Plane. Although a normal overhaul required 160 days, the task on the very first S2F was completed in only 124 days.

February saw VA-176 relocated to the station from NAS Cecil Field. It was then disestablished only to be re-established in June.

VP-18 began switching to jet-equipped P2V-7 Neptunes in April. The P2V-7 was powered by two jet and two reciprocating engines, making it the Navy's most advanced antisubmarine patrol plane.

On April 6, the $2.5 million helicopter hangar and repair shops began construction in the Overhaul and Repair area. The helicopter hangar was constructed on the site formerly used as the parachute loft.

The Overhaul and Repair shop was busy in May outfitting aircraft for the Byrd Expedition to the South Pole. R4D support aircraft and HO4S-3 helicopters were being fitted to withstand the rigors of the cold Antarctic weather. Included was a bright orange paint scheme!

Changes to the stations main gate were on tap during August. The circular islands, located at the entrance to the station, were removed. Further traffic improvements continued on August 29, as the West gate (Birmingham gate today) opened for traffic to enter and depart the station. No right turns were allowed during afternoon rush hour,

however, so "personnel could not beat the main gate traffic!"

The only aircraft ever lost by VW-4, Hurricane Hunters, occurred on September 26, 1955. Nine personnel and the aircraft were lost at sea, while tracking hurricane "Ione" in the Caribbean. The squadron had just earlier accepted its first NCN-121 Super Constellation aircraft in the middle of September. The Overhaul and Repair shops were also busy during September changing the aircraft color schemes, as the traditional Navy blue was replaced with task-oriented colors. In addition, the department started overhaul and repair on the F7U-3 Cutlass jet aircraft. The NAS Jacksonville Overhaul and Repair was designated as the overhaul point for all Cutlass jets used by the Navy.

At the same time that jets were causing a shift in Overhaul and Repair work, they also caused a shift in Air Operations. In October, the air tower had to be raised an additional story to provide a full view of the entire landing field and the air traffic patterns. The construction only took 45 days, and cost $44,890. The station celebrated its 15th anniversary with a visit from Secretary of the Navy, Charles S. Thomas. VP-3 was decommissioned in November, after many years of service, due to budget limitations.

1956

By January 1956, the Naval Air Technical Training Center had already graduated 80,000 students since recommissioning in 1951. The largest of the schools was the Airman School, which in five years graduated 54,647 Airmen followed by the Aviation Ordnanceman School with 9,168 graduates.

In February, a Fire Department Inspector who injured himself ended the station's all-time safety mark at 113 days without a mishap. The amazing 113-day record has not been matched since.

May 20 saw a huge air show and parade to celebrate Armed Forces Day. Capping off the air show was a simulated air strike of the base using 34 jets from Carrier Air Group Seventeen.

Heavy Attack Squadron 3 was formed in June, flying the A3D "Skywarrior" as part of the Navy's heavy attack program.

On August 5, the station was receiving the Secretary of the Navy's safety award. Also receiving an award

were Naval Hospital Jacksonville and the Naval Air Technical Training Center.

The huge helicopter repair hangar (where the P-3 Orions are reworked today), was completed in September, and the helicopter repair operation was transferred from Hangar 123 into the new 200x500 foot structure, designated as Hangar 101W. With the completion of this massive hangar, the Overhaul and Repair Department had roofed areas covering almost 24 acres. Thanks to this new facility, eight different types of helicopters could now be repaired, including the HSS-1, H04S-1/3, HR2S-1, HRS-2/3, HUL-1, HUP-2 and HTL-4/5/6. The R4D assembly line also moved into the new hangar.

An air traffic control center for joint use by the Navy, Air Force and the Civil Aeronautics Administration was approved in October. A two-story building was constructed next to the Air Operations Tower for use as the control center at a cost of $325,000. By the summer of 1957, all aircraft using northeast Florida air lanes were monitored from the control center. In November, the first Navy R4D modified during the previous summer by the Overhaul and Repair Department, became the first plane ever to land at the South Pole. Perhaps the most spectacular change made to this particular R4D was painting it a brilliant fluorescent orange color whose glow could be easily spotted over the vast white distances of Antarctica.

Painting was also completed on other aircraft at the Overhaul and Repair Department, with 18 different color schemes now being used. The once common Navy blue colors continued to be phased out while task-oriented colors became the rule. As a result, trainers were painted white and "International Orange", while aircraft used in areas of high air traffic density were coated with gray and white camouflage, and helicopters were decked out in orange, yellow-orange, gray, green or even a special fluorescent paint, each according to their assigned task.

As a result of its greatly expanded presence, the Navy began exerting considerable economic impact on Jacksonville by the end of 1956. NAS Jacksonville now had 10,000 military and 5,000 civilians with an annual payroll of $35,000,000. Cecil Field had three air groups, two air task groups, 300 civilians and a payroll of $8,000,000. Fortunately, the relationship between the City of Jacksonville and the Navy could not have been better, thanks to the combined efforts of all concerned.

1957

On January 17, an A4D "Skyhawk", assigned to VA-34, and an F8U "Crusader", assigned to VF-32, arrived on station and increased to six the number of station based planes that could break the sound barrier. The JAX Navy Federal Credit Union passed the $1 million mark in assets on 20 January.

In February, it was announced the Mayport would be the home for four carriers - USS Lake Champlain, USS Franklin Delano Roosevelt, USS Antietam and the USS Saratoga. Major changes to the airfield occurred in 1957 as parking ramps were added to the landplane hangars and a taxiway 1,231 feet long and 75 feet wide was installed.

On March 18, one of the more known names in history reported on board the station, as Lee Harvey Oswald arrived. He checked into the Naval Air Technical training Center for six weeks to attend Aviation Fundamental School. He would receive basic instruction in subjects such as basic radar theory, map reading and air traffic control procedures. He was required to deal with confidential material while in the latter course at the station. He received ratings of 4.7 in conduct and 4.5 in proficiency while at NAS Jacksonville, the highest marks he had ever attained! He would go on to have a date with history for the shooting of President John F. Kennedy, who also made a brief stop at the station while in the Navy.

By April, the Overhaul and Repair Department had begun rework on the world's largest helicopter, the Sikorsky HR2S, and on the smaller F8U "Crusader" jet.

Disaster struck NAS Jacksonville when a tornado hit the salvage yard one Saturday on June 6, 1957. Several buildings were destroyed and $21,000 in damages occurred but no lives were lost. President Eisenhower had just visited the station the day before. On June 28, Captain William S. Harris retired, leaving the station XO, CDR R. H. Smith as acting commanding officer.

Sadly, the last R4D "Skytrain" to be overhauled at the Overhaul and Repair was completed on June 14. Although the Overhaul and Repair had overhauled this aircraft off and on since 1945, it had eventually lost the project to private industry. Fortunately, the R4D workload was replaced by the S2F "Tracker" which, in turn, was replaced

a few short years later by the A3D "Skywarrior" and the F4D "Skyray", when the S2F overhaul program was moved to NAS Pensacola. The first A3D arrived on July 30, 1958.

Captain Ernest W. Parrish, Jr., became the fourteenth Commanding Officer, relieving Commander Smith on September 23, 1957.

In October, an Operations Department Maintenance Officer salvaged a $7 million Super Constellation at the station's airfield. The "Hurricane Hunters" aircraft had suffered an engine failure, which caused it to abort on takeoff. As the JAX Air News described the incident "The 63-ton aircraft wheeled off the runway to the right, traveled through a marsh for some 400 feet, then with a sickening jolt, dropped four feet down into a rain-swollen, mud-filled drainage ditch." Since the plane could not be moved forward or backward without causing further damage, Lieutenant Commander Willingham of the Operations Department had the plane lightened by some 50,000 pounds, thus allowing the front of the plane to be lifted. The ditch under the nose of the plane was then filled with gravel, and the crippled aircraft was moved back onto the runway without further damage. Finally, on December 15, the Navy Exchange Service Station was looted of $7,000. On December 24, two marines confessed to the robbery.

1958

1958 began with the dedication of yet another new building. The impressive half-million dollar Fleet Air Photographic Laboratory was dedicated on January 14, becoming one of the Navy's largest photographic laboratories. It was designed from the ground up to support fleet activities, particularly the NAS Jacksonville based photograph reconnaissance squadrons, VFP-62 and VAP-62. On January 20, an F4U Corsair flown by Lt Boyd of VA-104 crashed at Camp Blanding killing the pilot. This crashed aircraft was finally recovered by the Navy on May 20, 1991 and brought back to the station for disposal.

By January 23, the station was cutting civilian employees for the second time in six months. A total of 369 lost their jobs through Washington-mandated reductions.

As American's first satellite was launched into space in early February, communications workers at the station

were on the scene tracking the satellite, "Explorer." Although the satellite's signals, on a frequency of 108 megacycles, were only trackable for about three minutes, it was an exceptionally exciting time for the station personnel involved.

In March, Cecil Field based A4D crashed in Doctor's Inlet and divers and salvage crews from NAS Jacksonville were busy for a week recovering everything except the engine, which could not be found.

The Blue Angels were back to NAS Jacksonville in their new F-11F Tigers on April 13, and 90,000 fans came to witness the show.

On June 2, an FJ-1 Fury jet was given by the station to the City of Starke for placement in a park. Starke took advantage of a recent navy directive that allowed stations to give surplus aircraft to cities. In 1991, the author contacted the City of Starke to ask what ever happened to the plane. We were looking for old aircraft for our static aircraft park, and the FJ was now a rare aircraft. Much to my dismay, the next call the author received was from an investigative reporter from local Jacksonville NBC affiliate TV-12 asking why the Navy was checking into the aircraft. After a short explanation, the reporter stated he was friends with the Mayor of Starke, and was checking on his behalf. Starke did receive the aircraft and it did go into a park there. But, by the end of the first week, two children playing on the aircraft had already broken bones. An emergency meeting of the City Council was called, and the plane was sold for $201.27 to a recycler in Gainesville. The same plane today would bring approximately $1 million dollars, if available!

The station became involved with the space program again in July when VP-18 was the first to spot and track the nose cone from the Army's test firing of the first Jupiter C-1 rocket. The 22 of July had embarrassed Navy officials trying to explain how a practice rocket was accidentally fired into two houses in Bostwick, Florida (south of the station). No one was hurt in the accident.

Combat jet training was also in full swing in July, when another TV-2 jet trainer was assigned to the station. Besides keeping station pilots trained, the jets transported personnel to various east coast cities.

In August, LCDR R. E. Morris of VAH-1 became the first person to land an A3D Skywarrior with one wheel missing when he landed at the station.

The wheel broke off while practicing landings at Mayport, and he was diverted to NAS Jacksonville to make an emergency landing. Only one minor injury to one of his crew resulted.

Carrier Air Group 17 was abolished in September. On September 30, an A3D Skywarrior was having mechanical problems as it approached NAS Jacksonville. Three of the four personnel parachuted near the St. John's river. The pilot stayed with the aircraft as it flew just 400 feet over San Juan and Jammes streets. He then parachuted at a height of 375 feet. The plane crashed near the Hyde Park golf course, missing all houses. Souvenir hunters quickly descended upon the aircraft even as small explosions continued to go off. The Duval County Patrol showed up, and the aircraft was guarded for the rest of the night until removed by naval personnel. The pilot was credited with risking his own life to save others. By the end of December, it was announced the Naval Air Technical Training Center would be deactivated in 1959 for the second time.

1959

January 1959 saw three more squadrons scheduled for disestablishment to include Attack Squadron 105, Fighter Squadron 173, and the "Red Rippers" of Fighter Squadron 11. Fighter Squadron 43 was redesignated VF-11 and Heavy Attack Squadron One (VAH-1) moved to NAS Sanford, Florida.

An ultramodern "waterfall" spray painting facility opened in February in the Overhaul and Repair Department. This facility allowed for capture of paint vapors for increased safety of personnel and reduced air paint emissions.

The last S2F Tracker was completed in April, bringing the program to a close after reworking 205 of the aircraft since March 1955. The F4D program, which began on February 18 1958, replaced some of the work lost in the S2F.

Captain Parrish was relieved by Captain James Reedy, previously the Commanding Officer of the U.S.S. Lexington. Captain Parrish retired after 30 years of service and passed away in Jacksonville in 1995. (Note: I met Mrs. Parrish at a speaking engagement in 1992 and remained in contact with her until her death in 1996. She provided much information about their life at NAS Jacksonville.) In May, it was announced the Overhaul and Repair

Department would be operating under Navy Industrial Funds. This change meant that Overhaul and Repair Department would now operate as a business and "charge" its customers for work performed. This policy still exists today, as the Naval Aviation Depot continues to compete for work.

VP-18 was in the middle of the space program again in May when it spotted and tracked the world's first two astronauts, the space monkeys Alpha and Baker. After vectoring two destroyers and a Navy seagoing tug to the previously identified spot, the mission was successfully completed when the monkeys were recovered. This event was among America's first steps toward manned space flight under the Mercury Program.

In June, the station suffered its first major fire since being commissioned. Building 517 was totally destroyed and Navy and Jacksonville fire units were barely able to keep nearby structures from also going up in flames. The next day, the Naval Air Technical Training Center re-opened the Exchange in another building, with very little loss of service to personnel.

Also in June, lane control lights were installed on the now heavily traveled Yorktown Avenue in an attempt to control the increasingly hectic morning and evening rush-hour traffic. These lights were eventually replaced completely in April 1989, but the basic

traffic pattern established 30 years ago still remains in effect today. In addition to the lane control lights, the first traffic lights were also installed at two major intersections. Squadron activities included the disestablishment of Fleet Aircraft Service Squadrons (FASRon's) 6 and 109. In their place the Aircraft Maintenance Department was established to handle intermediate aircraft maintenance.

The first P3 Orion landed at the station on October 1, 1959. A Lockheed demonstration team brought in a P3V-1 to convince naval personnel to adopt this aircraft over the P2V Neptunes, which were then being used by the Patrol Squadrons. The scheme obviously worked, as a look at the airfield today will readily indicate.

The Blue Angels performed again at NAS Jacksonville on October 25 in front of an audience numbering in the tens of thousands. Meanwhile, the Overhaul and Repair Department was busy modifying a HUS-1 helicopter for the HIDAL (Helicopter Insecticide Dispersal Apparatus Liquid) program. The program consisted of modifying the helicopter for aerial pesticide spraying.

The decade of the 1950s again saw unparalleled growth for the station. This growth, and the station's vital importance, would be further shown during the Cuban Missile Crisis and the Viet Nam War during the 1960s.

A "Jax Air News" comic, making fun of activities at the Overhaul and Repair Department, published July 25, 1951. (Jax Air News)

1950 SQUADRON ACTIVITIES

The 1950's saw squadron activity at NAS Jacksonville the likes of which may never be seen again. Carrier Air Groups (CAG's) were very active at the start of the decade, and each had normally 3 or 4 fighter squadrons and two attack squadrons attached to them. Five CAG's were very active at the station consisting of CAG 1,4,8,13 and 17. To make tracking the movements of these CAG's and their accompanying squadrons more difficult, they routinely transferred back and forth from NAS Jacksonville to NAAS Cecil Field, and then to NAS Cecil Field when that station finally became a separate air station and no longer an auxiliary station under NAS Jacksonville. Squadrons were also changed from fighter to attack and visa versa. Additionally, a lot of squadron disestablishments and establishments occurred depending on the needs of the Korean War and the closing and reopening of Cecil Field. Some of the more significant squadron actions for the decade of the 1950's are detailed below:

VA-12 Nickname: Ubangis This squadron used the famous "Kiss of Death" insignia throughout the time it was stationed at NAS Jacksonville. It rotated back and forth no less than three times between NAS Jacksonville and NAS Cecil Field. On April 26, 1952, the squadron was embarked on the USS Wasp in the Mediterranean when the Wasp collided with the Hobson (DMS 26). The Hobson sank, taking with her all 176 crewmen. The squadron, flying F2H-2 Phantom jets at the time, had to wait on board the Wasp until repairs to her were completed in drydock at New York before the squadron could return home to NAS Jacksonville.

VA-15 Nickname: Valions VA-15 was assigned to NAS Jacksonville on January 9, 1950. On May 22 of that year, this squadron, along with all other squadrons in CAG-1, were designated as training squadrons. This squadron's primary mission was training of fleet pilots in attack aircraft. Training included glide bombing, dive-bombing, rocket firing, day-and-night tactics and carrier qualifications in the AD Skyraider.

VA-34 Nickname: Blue Blasters This squadron was stationed at NAS Jacksonville for approximately six months from Oct 1952 to February 1953. There were no significant events while the squadron was stationed at NAS Jacksonville. They flew the F2H Banshee jet while stationed here.

VA-35 Nickname: Black Panthers This squadron was assigned to NAS Jacksonville in 1958, flying the AD-5 Skyraider. There were no significant squadron actions in the 1950's, but from October - November 1962 it deployed during the Cuban Missile Crisis. On February 4, 1965 the squadron's

commanding officer was killed in an accident.

VA-36 Nickname: Roadrunners VA-36 was based at the station from July 1955 until April 1956. The squadron flew the F9F-5 Panther jet while stationed here.

VA-42 Nickname: Green Pawns This attack squadron was stationed at NAS Jacksonville from September 1950 until June 1951 and flew the F4U Corsair fighter. It would depart the station when transferred to Naval Air Auxiliary Station Oceana.

VA-44 (First Squadron) Nickname: Unknown VA-44 arrived at NAS Jacksonville from NAS Norfolk on February 12, 1949. It remained at the station until disestablished in June 1950 and flew the AM-1 Mauler.

VA-44 (Second Squadron) Nickname: Hornets This second VA-44 squadron was established at NAS Jacksonville September 1, 1950, and were then immediately transferred to NAS Cecil Field 18 days later. It would return to the station in October 1952 and would not be sent back to NAS Cecil Field until some 11 years later in February 1963. It flew a wide variety of aircraft while stationed here including the F4U Corsair, F2H Banshee jet, F9F Panther jet, A4D Skyhawk jet, and TV-2 jet. In June 1953, the squadron conducted its first operations, striking targets in Korea. During September - October 1957, the squadron saw it's role change from attack to fighter for a short period. In June 1958, the squadron's role was changed to a fleet replacement squadron. The new mission involved flight training for pilots and maintenance training for enlisted personnel. In 1963, the squadron was divided with half of it being redesignated as VA-45 and relocated to NAS Cecil Field.

VA-45 (First Squadron) Nickname: Fishawks
VA-45 had a short life at NAS Jacksonville. It arrived flying the AM-1 Mauler on February 14, 1949, and was disestablished on June 8, 1950.

VA-45 (Second Squadron) Nickname: Blackbirds
The second VA-45 was established at NAS Jacksonville on September 1, 1950. On April 10, 1951, their new insignia was approved which became one of the more well-known insignias of any Navy squadron as the "Four and Twenty Blackbirds." The squadron went to NAAS Cecil Field only 18 days after being established, but returned to NAS Jacksonville on October 12, 1952 to remain until disestablishment on March 1, 1958. They flew the AD Skyraider. In June 1953, the squadron flew its first combat operation while deployed to Korea aboard the USS Lake Champlain. This squadron flew 387 combat missions in Korea.

VA-66 Nickname: Waldomen
VA-66 was only stationed at NAS Jacksonville from April 1951 until September 1951, flying the F4U Corsair.

VA-85 Nickname: Black Falcons
This squadron was established as a reserve squadron at NAS Niagara Falls on February 1, 1951, and reassigned to NAS Jacksonville on April 5. The arrived flying the TBM-3 Avenger. They only remained at the station for six months, being reassigned to NAS Quonset Point in September.

VA-104 Nickname: Hell's Archers
VA-104 had two different tours at the station during the 1950's. The first was from April - December 1953, and the second was February 1957 until disestablishment on March 31, 1959. They flew the F4U Corsair for a short period, then changed to the AD Skyraider. In July 1958, the squadron operated from the USS Forrestal ready to enter the Mediterranean if needed to support the US Marines landing in Beirut, Lebanon.

VA-105 Nickname: Unknown
VA-105 also had two tours at the station. The first was from July 1955 until April 1956, and the second from November 1958 until disestablishment in February 1959. They flew the AD Skyraider for both assignments. When reassigned to NAS Jacksonville the second time in 1958, it's mission was changed from an attack squadron to training personnel in the AD-6 Skyraider for eventual assignment to fleet operations.

VA-106 Nickname: Gladiators
The Gladiators tour at the station lasted from October 13, 1952, until December 1954, when they were transferred to NAS Cecil Field. They flew the F2H Banshee jet.

VA-134 Nickname: Unknown
VA-134 had a short life at the station arriving in April 1949 and being disestablished in February 1950.

VA-172 Nickname: Blue Blots
VA-172 arrived from NAAS Cecil Field on March 24, 1950, and stayed until 22 February 1958. They flew the F2H Banshee and the A4D Skyhawk jets. This squadron, on August 23, 1951, participated in it's first combat sortie in Korea. It also marked the first use of the F2H in combat. Two days later, the squadron provided escort for 30 Air Force B-29 bombers in a raid at Rashin, North Korea. During November-December 1956, while attached to the USS Franklin D. Roosevelt, the squadron was ordered to deploy and operate off the coast of Spain as a result of the Suez Canal crisis.

VA-174 (Second Squadron) Nickname: Hell Razors
This squadrons original insignia was designed by Walt Disney in 1944, and remained the squadron insignia throughout their history. They were based at NAS Jacksonville from April 1, 1949, until April 1954. They flew the F4U Corsair and the F9F Cougar jet while assigned here.

VA-175 Nickname: Devil's Diplomats
VA-175 arrived at NAS Jacksonville on January 9, 1950 and remained at the station until disestablishment on March 15, 1958. They flew the AD Skyraider the entire time. In November-December 1956, they were ordered to deploy off the coast of Spain as a result of the Suez Canal Crisis, while on the USS Franklin D. Roosevelt.

VA-176 Nickname: Thunderbolts
VA-176 remained at the station longer than any other attack squadron, having been based here from February 1955 until transferred to NAS Oceana in May 1968. When the squadron left in May, it retired the last AD Skyraider in the Navy, and ended a long tradition of propeller driven attack aircraft for aircraft carriers. This squadron participated in the Suez War off the coast of Egypt in 1956; flew off of Central America aboard the Shangri-La and the Wasp to prevent Cuban infiltration in December 1960; operated off the coast of the Dominican Republic following the assassination of that country's leader in June 1961; operated from the Shangri-La during unrest in Haiti in 1963; was part of CVW-11, an all attack air wing and the first to deploy in Vietnam; conducted their first combat strikes in Vietnam on May 1966; became the first and only propeller driven aircraft to

shoot down a MIG-17 during an engagement on October 9, 1966, and lastly, when the Israeli forces attacked the USS Liberty on June 8, 1967, the squadron was ordered to proceed to the ship and defend it. Shortly after the aircraft were launched, the aircraft were recalled when Israel apologized for the attack.

VA-859 Nickname: Unknown VA-859 was a reserve squadron stationed at NAS Jacksonville in April 1951. No other history is known about this squadron.

VJ-62 Nickname: Tigers VJ-62 was based at NAS Jacksonville when established here on April 10, 1952, until October 20, when they were reassigned to NAAS Sanford, Florida. Their mission was aerial photographic intelligence. They returned to NAS Jacksonville, redesignated as Heavy Photographic Squadron SIXTY TWO (VAP-62) on August 15, 1957, and remained until they were disestablished on October 15, 1969.

VF-916 Nickname: Roaring Bulls VF-916 was a reserve squadron established at NAS Jacksonville on February 1, 1951. In September they transferred to NAS Oceana. Today, they are squadron VFA-83 flying the F/A-18 Hornet. When Cecil Field closed in 1999, the squadron left after 34 years of assignment there.

VF-921 Nickname: Sidewinders VF-921 was also a reserve fighter squadron initially established at NAS St. Louis, but actually formed when called to active duty at NAS Jacksonville on March 28, 1951. The squadron flew the F4U Corsair. On June 11, it transferred to NAS Oceana, then in March 1966 to NAS Cecil Field. The squadron flies the F/A-18 Hornet today and carries the designation VFA-86.

VAH-3

VA-165

One of the first rescue helicopters, an HUP-1, was assigned to the station in 1954. (U.S. Navy)

VF-17A jet leaves NAAS Cecil Field for NAS Jacksonville in February 1950. (U.S. Navy)

VA-36

Headquarters of Fleet Air Wing 11, which had just moved to the station in 1949. (U.S. Navy)

A Caroline Mars Seaplane anchors off the station in 1950. This seaplane was the largest to ever land in the St. John's River. (U.S. Navy)

CDR Ghiresquire, Commanding Officer of VP-3, admires the P2V model built by Petty Officer Burns, July 1, 1953. This photo appeared in the Jax Air News, and the model was apparently stolen shortly after that—never again to be located! (U.S. Navy)

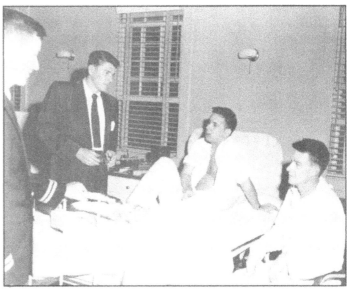

Movie star Ronald Reagan visits patients at Naval Hospital Jacksonville on January 20, 1951.

Helicopter repairs at the Overhaul and Repair Department, 1955. (U.S. Navy)

Capt. E.J. Mason, USMC (left) and LTG Axeil, USN, admire their shooting skills as the receive the Battle "E" Award while assigned to station squadron VF-12.

An HUP helicopter is reworked at the Overhaul and Repair Department on November 20, 1953. (National Archives #80-G-633275)

The first four pilots of the Red Rippers Squadron VF-11, return from Korea on March 23, 1953. (National Archives #80-G-482701)

The main administration building of Naval Hospital Jacksonville in 1954. (National Archives #80-G-686854)

Right: An HR2S-1, the largest helicopter when this picture was taken on April 5, 1957, shown after rework. (U.S. Navy)

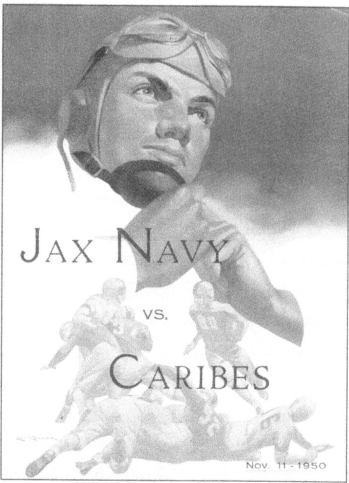

Football program for NAS Jacksonville "Flyer" Football Team. November 11, 1950.

This H04S-3S, painted orange, was reworked by the Overhaul and Repair Department in August 1956 for Operation Deep Freeze. (U.S. Navy)

The first jet ride for a woman at the station takes place on November 21, 1956. (U.S. Navy)

The mothball fleet at Green Cove Springs is shown here in this 1958 photograph. (National Archives #80-G-628947)

Naval Hospital Jacksonville, aerial view, January 13, 1956. (National Archives #80-G-686855)

The Blue Angels prepare for another air show, December 1952. (National Archives #80-G-476764)

The large crowd at the December 1952 air show. (National Archives #80-G-476765)

Engines are installed in this Air Force B-29 to correct a problem on July 27, 1951. (U.S. Navy)

A sailor conducting wheel watch, as a P2V of VP-18 approaches the station on January 12, 1955. (National Archives #80-G-658846)

Patrol Squadron 16 is born as VP-741 is redesignated and the logs are passed on February 16, 1953. (National Archives #80-G-478398)

The helicopter division of the Overhaul and Repair Department checks the blade balance on an HRS-1 (HO4S), September 26, 1952. (U.S. Navy)

RADM A.C. Read, the first Navy pilot to fly the Atlantic in NC-4, is helped into a jet by CDR Gilmore, Commanding Officer of Squadron FASRON 109 on October 21, 1953. (National Archives #80-G-629543)

Waving his arms, a NAS Jacksonville traffic cop directs the heavy outgoing 3:30 pm traffic down Albermarle Street from the Overhaul and Repair Department. (U.S. Navy)

The Commanding Officer

of

Light Photographic Squadron SIXTY-TWO

Cordially Invites You

To Attend

The Dedication

Of The

FLEET AIR PHOTOGRAPHIC LABORATORY

U. S. Naval Air Station

Jacksonville, Florida

on

Tuesday, 14 January 1958, at 1400

Reworking of an F4U Corsair Fighter

The following eight photographs show an F4U Corsair going through rework at the Overhaul and Repair shops in June 1951.

Left, top down. Corsairs scheduled for overhaul are received far right, then dismantled for inspection of all parts. Fuselage enters cleaning ramp area; paint is removed; internal compartments vacuumed and all oil and gas removed. In paint shop primer is applied. Engines built as quick change unit. Right, top down. Engines tested in test cell. Assembly line—engine installed with flight and engine instruments. Final stage—propeller installed and hydraulic system checked. Final painting. (U.S. Navy)

101—Philippine Mars over Jacksonville, Fla.

104—Officers' Club, U. S. Naval Air Station, Jacksonville, Fla.

102—Control Tower, U. S. Naval Air Station, Jacksonville, Fla.

VP-18 P2V-7 Neptune. (U.S. Navy)

P2V-2 Neptune is shown having the front gun loaded, July 23, 1951. (U.S. Navy)

AD Skyraider of VA-15. (U.S. Navy)

AD-6 Skyraider assigned to VA-45.

First Cougar Trainer, F9F-8T, assigned to station arrives April 16, 1957. (U.S. Navy)

An R-60 Constitution makes another delivery to the station.

A3D Skywarrior arrives at NAS Jacksonville, January 29, 1956. (U.S. Navy)

The CO and XO's cars match the squadron insignia, as this photo shows. (U.S. Navy)

AJ-2Ps assigned to photographic sqaudron VAP-62. (National Archives #80-G-449418)

F2H2 Banshee in front of the Air Operations tower on August 28, 1951. (U.S. Navy)

AD-6 Skyraider of VA-175. (National Archives #80-G-680408)

F7U Cutlass sits on the ramp, June 13, 1954. (U.S. Navy)

VA-175 receives a new AD-5 Skyraider on April 11, 1954. (U.S. Navy)

A Grumman JRF-5 sits on the ramp, January 30, 1952.

Above: F2H-2P heads straight up after takeoff from NAS Jacksonville, January 21, 1953. Aircraft was attached to the photographic squadron. (U.S. Navy)

Left top: P2V-2 in flight over station, July 3, 1953. (National Archives #80-G-484613)

Left: R2N-2F Banshee of station photographic squadron flies over station, December 1952.

This 1957 aerial photo shows 10 different aircraft:
1) F2H Banshee;
2) AD Skyraider;
3) HUP-2 Retreiver;
4) Bell Model 47;
5) HSS Seacat;
6) HUP-1;
7) S2A Tracker;
8) AF Guardian;
9) HRS-4 Chickasaw;
10) F-80 Shooting Star.
(U.S. Navy)

DEDICATION
ADM. JOHN TOWERS FIELD

20th Anniversary
COMMEMORATING THE COMMISSIONING OF
THE
JACKSONVILLE NAVAL AIR STATION
OCTOBER 14-15

ADMIRAL
JOHN HENRY TOWERS
U. S. NAVY
1885-1955
-PIONEER IN NAVAL AVIATION-
-WARTIME CHIEF BUREAU NAVAL
AERONAUTICS

1940 1960
JACKSONVILLE, FLORIDA

OFFICIAL CACHET: JACKSONVILLE STAMP COLLECTORS CLUB

OFFICIAL NAVY LEAGUE PROGRAM

25¢

25TH ANNIVERSARY
THE JACKSONVILLE NAVAL AIR STATION

The decade of the 1960's would find the station again involved with a foreign conflict, as the war in Vietnam would eventually involve Navy forces. There would be other trying times before that, however, as the Cuban Missile Crisis would bring the country to the brink of war! Squadron assignments would also take place during this decade, as the attack squadrons would eventually be lost from the station late in the decade. The Naval Air Rework Facility would also break away from the station and become a separate tenant command in 1967.

Patrol Squadron (VP) 30 was established on June 30, in a ceremony in Hangar 140. VP-30 joined VP-5, VP-16 and VP-18 under the command of Fleet Air Wing Eleven. That same day, Admiral Rickover was present at the station consulting with a Navy Study Group on Shipbuilding and Modernization in a top-secret meeting. At the end of the ten-day conference, the committee was to report to Admiral Arleigh Burke, Chief of Naval Operations. However, no official statements were ever given concerning the meetings, which were held under the strictest of security.

On July 23, lightning again struck the base, this time targeting a VP-30 P2V Neptune parked on the runway in front of Hangar 140. Fortunately, a sharp-eyed sentry spotted the fire, and, assisted by personnel from VP-16 and VW-4, extinguished the flames before they could cause major damage to the aircraft. Less than one month later, VW-4 (Hurricane Hunters) was transferred from NAS Jacksonville to Roosevelt Roads, Puerto Rico. July 29 saw a landing go terribly wrong for Heavy Photographic Squadron VAP-62. While landing on the USS Saratoga, the tailhook of their A3D-2P Skywarrior separated and the aircraft plunged off the deck into the sea taking with it the squadron commanding officer and two other squadron aircrewmen. It was never recovered.

Hurricane Donna hit Jacksonville in September, but caused little damage, thanks to the hurricane plans put into effect earlier.

On October 14-15, 1960, one of the biggest celebrations in the history of NAS Jacksonville took place in honor of its 20th Anniversary. After an inspiring evening concert by the Jacksonville Navy Band on the night of 13 October, a gala open house and top-notch air show were held starting the following day. Special invited guests included Secretary of the Navy William B. Franke and the station's first commanding officer, Admiral Charles P. Mason. The air shows started with a large flyover of vintage naval aircraft, followed by the ever-popular Blue Angels, and ended with a series of demonstrations by modern Jacksonville Navy aircraft. A major event during the air show was the naming of the airfield in honor of Admiral John Towers, pioneer of naval aviation and a longtime supporter of NAS Jacksonville. Mrs. Towers was at the ceremony that unveiled a plaque on the station airfield. A total of 100,000 people witnessed the air show and its spectacular display of aircraft. The show provided a fitting tribute to NAS Jacksonville since 60% of the Navy's Atlantic fleet air strike force was based at the station. The runways were the Navy's busiest, supporting 12,000 landings and takeoffs a month. As of the 20th Anniversary, the people of NAS Jacksonville had contributed over $750,000 to various local fund raising drives. For every dollar put up by the citizens of Jacksonville in 1939 to help the Navy locate the base here, at least $4,375 had come back to the community.

1961

1961 brought with it an unusually heavy burden of sorrow and loss for those at NAS Jacksonville. First, the base was stunned and saddened in January when a Military Air Transport Service C-118 mysteriously disappeared on a flight to Argentia, Newfoundland. The two Navy personnel and four civilians who were on board were never found; neither was the crash site. VP-30 lost eight more personnel in March as a P2V Neptune crashed and exploded in Lake George while on a routine training mission. Then, on May 20, another P2V Neptune crashed just off St. Augustine Beach, killing one. The aircraft that crashed, witnessed by hundreds of horrified bathers, was assigned to Patrol Squadron 742 of the Naval Air Reserve Training Unit. This crash was the first fatal accident since the "Weekend Warriors" were established in 1946.

The P5M Marlin flying boat arrived in July, assigned to VP-30. This marked the first time seaplanes had been at NAS Jacksonville since the end of WWII.

After an accident free June and July, the station's first traffic death occurred in August marring the 21-year death free traffic record held since the station was commissioned. The accident occurred when an insurance agent drove into the back of a bus, which had stopped, to discharge passengers on Yorktown Avenue at Gillis Street. Mr. Narron was pronounced dead at the scene. Since his death twenty-nine years ago, three more fatalities have occurred through January 2000. The second fatality occurred in 1970, as a young newly promoted officer left his celebration at the Officers' Club and drove head on into a large oak tree located just 200 yards from the club. Considering the millions of cars and tens of thousands of personnel that transverse the base, it's not a bad record.

August saw the establishment of VA-135 at the station. The squadron would only remain at the station until July 1962, when it would transfer to NAS Cecil Field.

Captain James R. Compton became the 16th Commanding Officer on September 25th. The informal change of command was held in the office of the Commanding Officer. On October 15, another large air show and open house were conducted in honor of the 50th Anniversary of Naval Aviation. At the same time, VP-7 and VP-741 were added to Fleet Air Wing Eleven. VP-7 was transferred from NAS Brunswick to NAS Jacksonville to assist in the Project Mercury Space Program, and VP-471 was originally established in response to the Berlin crisis and then to support operations during the Korean War.

1962

As was the case in the previous year, 1962 began with a loss when a P2V Neptune from VP-5 disappeared off of Iceland on January 12, 1962. Like the C-118 before it, the plane was never found. A memorial service was held in the All

Saints Chapel in early March for the crewmembers.

In April, the Naval Air Technical Training Unit (whose name was recently changed from the Naval Air Technical Training Center), opened the Navy's first Radiography School. The eight-week course was designed to equip Navy personnel to use radiation in detecting structural defects in Navy aircraft. Also in April, a fire severely damaged the Enlisted Mens' Club "Bluejackets Inn." The club was closed until August 14th while repairs were made.

In May, VP-18 was in the headlines as the first crew to spot Astronaut Scott Carpenter. The plane spotted the capsule containing America's second spaceman after nearly three-quarters of an hour had passed without the nation and the world knowing if the astronaut had safely splashed down. The VP-18 crewmembers were invited to be special guests on the popular television show "What's My Line," and managed to thoroughly stump the panel. Also in May, VP-30 commissioned Detachment "A" for the purpose of training personnel to operate the new "P3V-1 Orion."

The Overhaul and Repair (O&R) Department, busy working on helicopters, added a new metal forming press in August. The press could provide pressures of 5,000 pounds per square inch and became the largest piece of operative production machinery at the station. Later, the O&R Department underwent a major change on October 1 as it embarked under a new concept of operation called Navy Industrial Fund (NIF). This program, which is in effect today, required the O&R to charge for work and services performed. The O&R was the largest of the 13 departments operating at NAS Jacksonville in 1962, with a total of 3,077 employees.

October and November were trying times for the station. The Cuban Missile Crisis was all the station was concentrating on. Squadron VA-35, the "Black Panthers", were ordered to go to McCalla Field, Guantanamo Bay, Cuba. The squadron remained there for two months ready to provide any support as needed in an attack mission. The photographic squadrons were busy taking intelligence photographs that were flown back to the station, and immediately processed at the Photographic Building. After this, the photographs were then immediately flown to Washington where the President had them just a short few hours after processing. The station's patrol squadrons were providing constant surveillance activities near Cuba. Other aircraft at the station remained on a high alert, armed and ready for any action necessary. On October 25, the station was trying to quash reports that Navy dependents in the Jacksonville area were being evacuated due to the growing crisis. Fortunately, the situation was resolved and the station secured from its high alert back to a more day-to-day business posture. On October 16, the flying club was organized for the purpose of providing a form of off-duty recreational flying, and still exists today. The club was initially based at Herlong Field.

1962 ended as the first P3A Orion was delivered to VP-30. The switch to the Orion by the other squadrons would not take place until early 1964.

1963

In 1963, the Navy received the UH-2A Seasprite, and the O&R Department was named the prime rework site for this helicopter. The UH-2A was designed to be able to take off and land from a moving ship at sea under any weather conditions. The Overhaul and Repair facility was also the prime rework site for the CH-46 and CH-53 helicopters.

Once again, as during the previous two Januarys, a tragic plane crash occurred, but this time there was a happier ending. Another P2V Neptune, this one from VP-30, ditched in the cold Atlantic about eight miles off Jacksonville Beach. Luckily, the ten men on board were all rescued from the Atlantic. One of the NAS Jacksonville rescue helicopters had to make an emergency landing itself at Spring Glen Elementary School after developing engine problems on the return trip back to the station. In February, NAS Jacksonville based VA-44 was broken up into two separate training squadrons. The part of VA-44 which specialized in jet pilot training and was comprised of 56 jets and 500 officers and men, was transferred to Cecil Field. This move was partly designed to increase efficiency by placing all jet training personnel, planes and equipment at one locality. However, the main reason for the move was to reduce community complaints about the jet noise around NAS Jacksonville. The non-jet part of VA-44 was recommissioned as a new squadron, VA-45, and remained at NAS Jacksonville. The new squadron operated about 25 propeller aircraft including the A-1H Skyraider bomber and the T-28 Training Plane.

The present NAS Jacksonville insignia was also accepted in February. It was changed only slightly from the design submitted by a JAX Air News contest winner. This marked the fourth insignia for the station since its inception. The station had not had an insignia since WWII ended.

In July, the 300-unit Dewey Park housing area was finally closed. Concerns about noise and safety resulting from the housing area being located in the landing and takeoff zone prompted the closing.

The station's second murder was the talk of the station on October 12 as a wife fatally shot her husband in the parking lot of the Acey-Duecy (Enlisted Man's) Club on station. She waited at the club until security arrived and handed her gun over to them. Her husband died 20 minutes after the shooting at the Naval Hospital. She was later found guilty of first-degree murder.

On December 29, 1963, another fire broke out, this time in the Hotel Roosevelt in downtown Jacksonville which eventually killed 21. When the lower floors became encased in flames, a number of persons had made their way to the roof of the 13-floor hotel in a desperate attempt to reach safety. Once there, they were trapped and the Navy was called to rescue them. Four helicopters from NAS Jacksonville and Naval Air Reserve Training Unit Jacksonville, along with one helicopter from NAS Cecil Field, answered the call for help. A total of 14 fortunate souls were saved, 11 by the NAS Jacksonville helicopters and 3 by the NAS Cecil Field helicopter. The pilots had to use extreme skill to avoid the roof air conditioning units. They dared not actually land on the roof, as they had no way of knowing whether the roof was designed to hold the weight of the helicopter. Numerous letters of appreciation and Air and Navy Commendation Medals were delivered to the Navy personnel who saved the 14 lives. The station finished 1963 by receiving the Secretary of the Navy Award for Achievement in Ground Safety.

1964

January 1964 saw a VP-18 P2V Neptune making a wheels-up landing on the NAS runway. Luckily, none of the personnel in the seven-man crew was seriously hurt, other than their pride!

In February, President Johnson landed at the station. Despite the pouring rain, the President read a short speech, then went into the crowd to shake hands. A few minutes later, he boarded a Marine helicopter for Palatka where he set off a dynamite charge, beginning construction of the Cross-Florida Barge Canal.

In March, the O&R Department, was finishing the overhaul of the first A-5A "Vigilante," after recently being designated as the overhaul point for this aircraft.

VP-45, who had just recently arrived at the station from Bermuda, was the first station operational squadron to receive the P3A Orion. This plane, a military adaptation of the commercially popular Lockheed Electra, was designed to replace the aging fleet of P2V Neptunes. A total of nine Orions were eventually to be issued to each squadron. Although additional modifications and updates have continued on this aircraft, it will probably remain on active service with the squadrons at NAS Jacksonville well into the next millenium. The first operational flight for the Orion took place on March 13.

On April 23, Mrs. Barbara Hennessy, the first baby born at Naval Hospital Jacksonville in December 1944, gave birth to her first baby there. This was the 26,277th baby delivered at the hospital since it opened 19 1/2 years earlier. Also in April, the runway was closed for ten days while new aircraft landing arresting gear was installed. During the installation period, all of the air station's aircraft used Cecil Field's runways.

In June, the O&R Department had its first fatality. A helicopter engine was being tested inside an O&R building when it apparently malfunctioned and began to self-destruct. As the blades came loose and started flying in every direction, one of the projectiles pierced the side of a nearby helicopter, mortally wounding the unfortunate employee in the chest and neck. After five hours on the operating table, the man died.

On June 30th, Captain John E. Mackroth relieved Captain Compton to become the air station's 17th Commanding Officer. The two captains had previously served together in Korea and also had been classmates at the Naval War College.

President Johnson made his second trip to NAS Jacksonville in September to view the damage caused by Hurricane Dora. NAS Jacksonville was relatively lucky, sustaining about $193,000 in damages compared to Duval county's millions of dollars in losses.

Squadron VU-10 (Utility Squadron TEN) Detachment was disestablished on July 14. The squadron mostly flew the B26 and only recently was it replaced by the US2C. This was the squadron that towed targets for fighter gunnery practice. They left the station with a perfect nine-year safety record.

November 1 saw NAS Jacksonville celebrating its 24th anniversary with an air show featuring the Blue Angels. Another highlight of the show was Dick Schram, the "Flying Professor."

In December, an aircraft from Heavy Photographic Squadron 62 made a mercy flight to Costa Rica to carry coral snake antivenin to a small boy bitten by a snake. The boy's life was eventually saved by the trip.

On December 22, 1964, NARF phased out the A-3 Sky Warrior rework program after completing 385 aircraft. The last aircraft departed the station for Rota, Spain. The program was replaced by the earlier arrival of the A-5 Vigilante. On December 25, a fire broke out at the Acey-Duecy club, but only caused minor damage.

1965

On January 21, 1965, the VW-4 Hurricane Hunters and their Super Constellations returned to NAS Jacksonville. They had previously been stationed at NAS Jacksonville from their commissioning in 1952 until 1961, when they were transferred to Roosevelt Roads, Puerto Rico. Since the Fleet Weather Facility had been moved to the station in 1964, the two units could now work even more effectively together.

On March 6th, the "Dawdling Dromedary," a Sikorsky SH-3A Helicopter, set an unofficial world distance record by flying the 1,840 miles from San Diego to Mayport, nonstop. This historic flight shattered the previous world record by 1,000 miles.

The CPO Club, at the corner of Yorktown and McFarland was destroyed by fire in May. This was the first major fire in five years.

The first major change in aircraft color schemes occurred in June. During combined P3A Orion and submarine exercises conducted the previous year, the submarine commander reported that the dark Seaplane Gray paint used on the Orion was easily spotted in the sky. He added that the lighter Gull Grey blended in well with the sky and was very difficult to see. As a result of these observations, the underside of the P3A Orion was painted with Gull Grey, while the topside remained in Seaplane Gray. This combination of light bellies and dark backs continued until 1988, when the P3s (and all other operational Naval aircraft), were ordered painted totally in Compass Ghost Grey. This revised tactical paint scheme (designed to absorb radar) was completed on all aircraft by 1992.

In August, the O&R Department assisted in "cocooning" 300 aircraft for overseas shipment. The process consisted of spraying aircraft with several coats of plastic compound, which then dried to form a covering similar to white plastic wrap. The aircraft could then be loaded onto the open decks of ships at Mayport and transported to Vietnam.

The station marked its 25th anniversary in October with an air show featuring the Blue Angels. Almost 250,000 spectators came to join in this gala Silver Anniversary celebration.

1966

VP-30 left Jacksonville in January 1966 for its new home base at Patuxent River, Maryland. At the same time, the squadron was converting to the P-3A Orion aircraft, having already transitioned five operational squadrons to the Orion.

On February 14th, the O&R Department was designated as a rework site for the A-4 Skyhawk. At the same time the department announced that rework would commence on the A7 Corsair II in September. The O&R Department was now second only to NASA at Cape Canaveral as the largest military industrial activity in Florida.

April saw the first station based squadron leave for Vietnam as VA-176 "Thunderbolts" headed off on the USS Intrepid. The squadron would go on to fly 1600 sorties against enemy positions. The highlight of the missions came when LTJG Tom Patton shot down a MIG jet in his prop-driven plane! Returning on November 19 with the squadron was Cecil Field based squadron VA-15.

The station runway was again closed during June and July while centerline lighting and east approach lights were installed. On July 4th, one wing of the CPO Barracks Building 700, including all personal belongings, was destroyed by fire. Ten days later, ground was broken for the new $12 million dollar VP Complex, Hangar 1000. The design was revolutionary for the Navy in that, for the first time, the squadrons would be colocated with the Supply and Maintenance functions of the Aircraft Maintenance Department. The complex had been the brainchild of a few dedicated officers in Fleet Air Wing Eleven in August 1963, and their idea had moved from concept to concrete at a phenomenal rate.

Captain John Stack relieved Captain Mackroth on July 22nd. Captain Stack was already familiar with NAS Jacksonville, as he had been stationed at various commands here during 1943-1947. On the Saturday before the Change of Command, a C-119 took off from the station and suffered an engine fire almost immediately after leaving the runway. At seven thousand feet, all 34 personnel donned their parachutes and jumped, some for the first time. Three minutes later, the empty plane crashed near Otis Road and the Atlantic Coast Line Railroad tracks in Jacksonville. It then took rescue teams many hours to find all of the men dangling in trees by their parachutes.

VP-5 retired the last P2V-5 Neptune on August 5. VP-5 was the last of the operational squadrons to transfer to the P-3. The P2V "Neptune" had served the patrol community well since 1948 when it was first introduced. From then on, all VP squadrons at NAS Jacksonville, except reserve squadron VP-62, operated the P3A Orion.

November saw the demolition of the NAS Jacksonville Band Shell. The twenty-four year old frame structure, located outside of the Administration Building, had outlived its usefulness and purpose, and was torn down to make way for a landscaped park. During WWII, the station band would play for personnel every morning after colors and often during noon hours, when an appreciative audience would sit around the bandshell, listen to the music and eat their lunches.

A tragic crash of an NAS Jacksonville reserve helicopter also occurred in November. An SH34J helicopter flying near Old St. Augustine Road developed mechanical trouble. The pilot made an emergency landing at a nearby field. Mechanics were flown over by a second helicopter to make repairs. After a five-hour repair job and

some ground testing the craft took off for the return trip to NAS Jacksonville. The craft rose to about 500 feet and "then just fell like a rock." The pilot must have known it was hopeless because he began working the helicopter away from the crowded residential section. It crashed in a residential neighborhood after striking an oak tree and exploded. The mechanic flown out to repair the helicopter was killed and two Navy reservists were badly burned.

The Department of Defense transferred part of Heavy Photographic Squadron VAP-62 from NAS Jacksonville to Albany, Georgia, in December as part of a planned cutback. NAS Sanford was also scheduled for closing. Meanwhile, NAS Cecil Field received an increase in manpower, and despite budget constraints, ground was broken at NAS Jacksonville for a new $7 million hospital. The hospital was designed to serve the estimated 122,287 military personnel, dependents and reservists located in the area.

1967

The NAS Chaplain's Office was instrumental in bringing an end to the ban on motorcycles by late March by starting a 90-day test period to allow motorcycles to enter the station. The cycles had previously been used on station by Marine Sentry personnel, but then had been banned from the station by late 1941. Several early station pioneers had told me the story that a motorcycle rider came back to the station late one evening in 1941 and hit a tree along the roadway. The tree was immediately cut down, and motorcycle riding was banned. Except for a brief period in 1956, the ban remained in effect until finally deleted for good on June 28. The day before this, a

27-stall riding stable was opened on station. Also in March, the Navy established a Family Service Center program that officially opened at NAS Jacksonville on the 17th. The service was established as a result of a recommendation by the Secretary of the Navy on Personnel Retention, which identified the need to assist new arrivals and those personnel with special problems. From its humble beginnings, the Service Center has expanded to include numerous beneficial programs servicing thousands of personnel today.

On April 1, the O&R Department was redesignated the Naval Air Rework Facility (NARF). Along with the

redesignation, NARF became a tenant activity, no longer a department of NAS Jacksonville.

The Candlelight Club, located where the 19th Hole Snack Bar and Golf Pro Shop is located today, opened on April 28. Originally, the club was designed to cater to parties, special events, celebrations and dinners of a more "formal" nature but after 10 relatively unsuccessful years, the club was closed and converted to a golf store and social area.

Later in September, the station's new communication center, Building 506, was nearing completion, as NARF started processing the first A-7A Corsair II. In December 1965, NARF had been designated as the prime rework activity for the A-7A. The A-7 served as the prototype for PAR (Progressive Aircraft Rework), processing. The A-7 remained a prime rework aircraft for the Naval Aviation Depot (formerly the NARF) until 1997. The first TA-4F was inducted on September 6 and was completed by December 13.

In November, the base suffered another fire when the "Waves" barracks (building 710) was damaged. A smoldering cigarette started the fire.

December was marked by two major events. First, on December 7, a 16-year era came to an end with the last test flight of the 2,123th helicopter to be overhauled at NARF. The helicopter program was transferred to NAS Pensacola. During the previous 16 years, over 20 different models had been overhauled, ranging in size from the small two seat TH-13 Sioux to the huge CH-37 Mojave, capable of carrying 35 armed Marines into combat. NARF mechanics had traveled the globe from frigid Iceland to the jungles of Viet Nam to keep the helicopters flying in support of the fleet. When the helicopter overhaul program began in 1959, NARF became the first facility designed to meet the specific requirements of this program. NARF next turned its attention to overhaul of the A4 Skyhawk, the A5 Vigilante and the A7 Corsair II. Along with the program change came a new challenge for the station; controlling jet noise.

The second major event was the dedication of the Naval Hospital's new facility on December 9. Representative Charles Bennett delivered the dedicatory address of the new eight-story hospital before military and civilian dignitaries from Washington, D.C. and the Jacksonville area. The new complex, situated on two acres, replaced

49 buildings that had been spread over 50 acres. By July 24, 1968, the 49 wooden structures which served as the hospital complex would all be demolished. All construction for the Navy Hospital was completed and the hospital was ready for patients on 15 December. (An interesting note to this demolition was discovered in January 2000. A parking lot addition was being constructed east of the present-day hospital, and work came to a halt when it was discovered all of the concrete slabs, wires and piping remained from the original hospital buildings. The contract was stopped for three months while negotiations with the contractor were conducted to remove the remaining parts of the original hospital buildings.)

1968

On April 25, NAS Jacksonville based VA-176 became the last squadron in the Navy to retire the A-1 Skyraider. (This was not even recognized by the Naval Historical community until 1996, having given credit to a California squadron as being the last to retire the Skyraider.) This also brought to an end the era of Navy propeller driven attack aircraft. Never again would they be seen at any air station or on an aircraft carrier.

The history of the Skyraider goes back to 1944 when the Navy issued an urgent call to Douglas Aircraft for a plane to replace the SBD Dauntless, a workhorse of the Navy's air arm during WWII. Working through a single night in a Washington Hotel room, three Douglas engineers came up with the basic design for the aircraft. The first production models were delivered in June 1945, and from that time until the final plane was retired at NAS Jacksonville, the Skyraider was considered the most powerful single engine propeller driven plane in existence. On May 3, a week after the final Skyraider was retired, VA-176 changed its home base to NAS Oceana, Virginia. The squadron had been at NAS Jacksonville since its commissioning in 1955.

May 3, 1968, also saw the dedication of Patrol Squadron Hangar 1000. VP-7 was the first squadron to move into the hangar. Although the hangar was only designed to hold two P3 Orions per segment, squadron personnel figured out a way to fit three when necessary. Today, the facility is almost always full of aircraft but its mission has changed slightly. The patrol squadrons only have

three of the five segments. The S-3 Viking squadrons which transferred from Cecil Field occupy the other two segments. The last P-5 Marlin in the Navy landed at NAS Jacksonville on July 9, enroute to the National Armed Forces Museum, a part of the Smithsonian Institute, to be put on permanent display. As the P-5 departed, the station's 19th Commanding Officer arrived by boat. Captain H. O. Cutler docked his sailboat, the "Scotch Mist," at the marina after a 12-day rough-weather trip from Pasadena, Maryland, thus becoming the first Commanding Officer to arrive under sail.

In September, plans for the new $443,455 Chief Petty Officers' Club were announced.

October 1st was designated as Consolidation Day for Jacksonville and Duval County as the city and the county merged into the largest area city in the U.S. NAS Jacksonville participated in the celebration by providing a band, a saluting battery and a Navy Color Guard. Captain Cutler represented the Navy at the official ceremony by placing Navy memorabilia in a time capsule that was buried behind City Hall. The capsule is scheduled to be unearthed in the year 2000. Patrol Squadron 18 was also disestablished the same day. The squadron had been based at NAS Jacksonville since 1950.

For ten days in December, fire-fighting experts gathered at NAS to test the revolutionary "light water" fire-fighting agent, so named because of its ability to make water float on flammable fuels, which are lighter than water. Both fresh water and salt water was used in the tests, which simulated aircraft fuel on a flight deck of a ship. The results of the test proved that rescue teams could move into hot areas and fight major fuel fires only 15 seconds after ignition.

1969

Between 6:00 P.M. on the evening of January 8 and 9:00 A.M. on January 9, $363,051.13 was stolen from the Barnett First National Bank on base. The thieves entered through a window, broke into the safe and were gone. According to a long-standing rumor, a Security Officer actually drove past the pair as they walked down Birmingham Avenue in the early morning hours of January 9, both carrying unusually large bags. Although the officer thought it was a little strange for the pair to be carrying such large bags at that time of morning,

particularly as they were travelling in the opposite direction of the barracks, he chose not to pursue the matter and the thieves made their getaway. However, their luck ran out in February when the FBI in San Antonio caught the pair. Eventually $284,000 of the stolen funds was recovered.

In March, Lieutenant Commander Coll E. Robertson was flying a UH-2A Seasprite helicopter over Orange Park heading toward Cecil Field when he "heard a loud bang and lost power," over a populated area of Orange Park. Robertson skillfully managed to crash the helicopter between two houses, resulting in only minor injuries to the helicopter crew and little damage to private property. Witnesses said it would have been very easy for the pilot to land on a house, but "he kept [the helicopter] up long enough to clear them."

In May, a P2V Neptune attached to Utility Squadron 8, at Roosevelt Roads, Puerto Rico, crashed into the St. Johns River while attempting to land at the air station. One of the eleven men on board was killed. Originally, the plane appeared to be intact, but as a barge lifted it out of the St. Johns for transport to NARF for salvaging, the port wing broke and fell off.

Hurricane Camille pounded Louisiana, Mississippi and Alabama in August. VP-45 and VP-16 flew tons of badly needed supplies to the disaster stricken victims. At the same time, the "Hurricane Hunters" Squadron was taking a beating in the press. Dr. Robert Simpson of the National Hurricane Center in Miami criticized the NAS Jacksonville-based crew for not penetrating Camille's eye to provide more accurate information of the storm's fury. Chief Warrant Officer Ray Boylan, the "Hurricane Hunters" information Officer (who would later become the meteorologist at a local TV station for many years), defended the crew's decision not to penetrate the storm. The maximum storm winds allowed for the "Super Constellation" plane flown by the "Hurricane Hunters" was 137 miles per hour. Since Camille's terrible power far exceeded that limit, the aircrew prudently decided not to violate the safety criteria.

President Nixon was trying to fight inflation in mid-1969. As part of his cost cutting efforts, two NAS Jacksonville based squadrons, VAP-62 and VP-7, were disestablished, along with two squadrons based at Cecil Field, VF-13 and VF-62.

The Accy-Deucy Club suffered a fire in October, sustaining $20,000 in damages. On October 25th, a Naval Reserve Aviator was killed as his jet crashed into the St. Johns River on approach to NAS Jacksonville. No other major event occurred to the end of the 1960s, and the station personnel continued to support the mission in Viet Nam.

As the 1970s approached, NAS Jacksonville's appearance would change somewhat with the arrival of the helicopter squadrons and their Wing.

1961 Open House Brochure. All images this page U.S. Navy.

24th Anniversary Celebration

Open House   1 Nov. 1964

CHANGE OF COMMAND CEREMONY

COMMANDER LEWIS IRVIN WOOD, USN

TO BE RELIEVED BY

CAPTAIN DONALD HENRY HEILE, USN

AS COMMANDING OFFICER

TEN O'CLOCK SEPTEMBER 26, 1969
JACKSONVILLE, FLORIDA

Squadron Disestablishments of the 1960's

Patrol Squadron 18 (VP-18) Left NAS Jacksonville April 1, 1965

Patrol Squadron 18 traces its history back to the Korean was and the activation of Patrol Squadron 861. VP-861 was the first reserve Patrol Squadron to be called to active duty during the Korean conflict. At that time, the fall of 1950, VP-861 was attached to the Naval Air Reserve Training Unit at NAS Jacksonville, but was based at Naval Air Station Norfolk, Virginia. In January 1953, the squadron was redesignated VP-18 (Flying Phantoms) and the home-base was changed to NAS Jacksonville.

As soon as the squadron came to NAS Jacksonville, the first order of business was to fly to P2V aircraft around the world to acquaint our allies with the new Patrol Bomber. It was on October 27, 1953 that a VP-18 aircraft was attempting to be the first to fly direct to Iceland from the station. The aircraft and all of the personnel on board was lost, with no trace of anything ever found. A memorial service was held at the station on October 27, 1953. In April 1954, VP-18 began a five-month deployment at Naval Station Argentia, Newfoundland. During this time, the squadron became the first P2V-5 and the first Fleet Air Wing Eleven plane to fly over the North Pole. In March 1955, the squadron became the first to receive the new jet-equipped P2V-7 model aircraft. In October 1955, at the request of the Icelandic government, the squadron depth-charged and conducted machine gun attacks on killer whales, which were destroying the herring schools. ADM Arleigh Burke, Chief of Naval Operations, commended the squadron in November 1957 for their part in the successful recovery of an Army "Jupiter" missile nose cone. This cone later appeared on a nationwide telecast as President Eisenhower explained how the U.S. had solved the space re-entry problem. In May 1958, the squadron was again involved in the space program by assisting in the successful recovery of the fourth "Jupiter" nose cone, carrying the first living things rocketed into space. These were the monkeys Able and baker.

The squadron continued their work with the space program as they got the first visual sighting of Astronaut LCDR Scott Carpenter on May 29, 1962. In October 1962, crews from VP-18 assisted in the Cuban quarantine during the Cuban Missile Crisis.

On April 1, 1965 the squadron officially shifted their home-port to Naval Station Roosevelt Roads, Puerto Rico. They were later disestablished there on October 1, 1968.

Patrol Squadron Ten (VP-7) Disestablished October 7, 1969

Patrol Squadron Seven, nicknamed the "Falcons," was established as VP-119 in August 1944 at Camp Kerney, California. It served in WWII and remained in the Far East until 1947. The squadron was renamed VP-ML-7 at that time and was relocated to MCAS Miramar for transition to the P2V Neptune. The squadron shifted home-port moving to NAS Quonset point, Rhode Island. In September 1948, the squadron was redesignated Patrol Squadron Seven.

In 1953 the squadron again shifted operations to the pacific, becoming the first East Coast patrol squadron to join in the Korean Conflict. It remained there until 1954, when they returned home to Quonset Point. Patrol Squadron 7 transitioned to the SP-2E aircraft in 1955, and in 1956 changed home-port to NAS Brunswick, Maine. For the next seven years, VP-7 operated out of NAS Brunswick.

In September 1961, the home-port of VP-7 was relocated to NAS Jacksonville, Florida and operational and administrative control was shifted to Commander, Fleet Air Wing Eleven. This was the eight, and last move for the squadron. VP-7 participated in the quarantine of Cuba during the Cuban Missile Crisis in October 1962, along with other NAS Jacksonville based squadrons. December 1964 saw the squadron transitioning to the SP-2H jet-equipped Neptune. They then had a busy 1965 with operations in Guantanamo Bay, Cuba, and NAF Sigonella, Sicily.

In early 1969 rumors of disestablishment began to circulate as the squadron was being eliminated as part of defense cuts. The squadron also received their first P-3 Orion on February 23, 1969, as they began to phase out the old SP-2H aircraft. They were disestablished at NAS Jacksonville on October 7, 1969, along with VA-106 and VA-64 at NAS Cecil Field.

Heavy Photographic Squadron 62 (VAP-62) Disestablished October 15, 1969

Heavy Photographic Squadron 62, nicknamed the "Tigers," was established as Photographic Squadron 62 (VJ-62) on April 10, 1952 at NAS Jacksonville. They were established initially flying the P4Y-1P patrol bomber. They phased this aircraft out in September 1952 and commenced flying the AJ-2P. The squadron's mission was aerial photographic intelligence for naval operations, as directed by Commander, Fleet Air Jacksonville. On October 20, 1952, the squadron changed home-port to Naval Air Auxiliary Station Sanford, Florida. The remained there until July 1955, when they again changed home-port to NAS Norfolk, Virginia. On August 15, 1957 the squadron changed home-port for the last time, as they returned to NAS Jacksonville. On October 19, 1959, the squadron transitioned to the A3D Skywarrior.

On July 29, 1960, the squadron lost its Commanding Officer along with two crewmembers while attempting to land on the USS Saratoga. The tailhook of their A3D-2P separated and the aircraft plunged off of the deck into the sea. This aircraft was the last jet not to have ejection seats, which proved fatal to this crew. October 1966 saw the squadron transfer a detachment of aircraft and personnel to VAP-61 to augment that squadron's operations in Vietnam.

The squadron was disestablished on October 15, 1969, at NAS Jacksonville as the fourth Jacksonville-based squadron eliminated during the defense cuts. They had conducted special photographic projects in Saudia Arabia, Turkey, Italy, Spain, various countries in northern Europe, Greenland, Iceland, Labrador, various Central American and Caribbean countries, Morocco, the Mediterranean and Atlantic Ocean areas and numerous places in the United States.

Fires were numerous in the 1960s as the CPO Club burned in May 1965; the BOQ in March 1967, and the Acey-Deucey Enlisted Club burned in October 1969. (U.S. Navy)

German officer and enlisted personnel receive a certificate from Commanding Officer, Patrol Squadron 30 upon completion of an anti-submarine warfare course, December 1965. (U.S. Navy)

President Johnson and his wife arrive at the station in February 1964. They were to go to the commencement ceremonies to start the Cross Florida Barge Canal. (U.S. Navy)

NAS Jacksonville helicopter rescues victims from the roof of the Hotel Roosevelt, which was in flames, December 29, 1963. (U.S. Navy)

Above: President Johnson meets the personnel assembled at the station in a rainstorm, February 1964. (U.S. Navy)

Left: Main gate, 1960. Motorcycles were banned from the station and are shown parked on the left. (U.S. Navy)

This non-flyable P2V Neptune, being hoisted aboard a barge at NAS Jacksonville, is on its way to Norfolk, VA for a complete overhaul, instead of the graveyard. It was prepared for shipment by the Overhaul and Repair Department. This effort, conducted in July 1962, saved the $2.5 million aircraft. (U.S. Navy)

A RA-3B Skywarrior of VAP-62 on the ramp on October 1, 1962.

Secretary of the Navy, Paul Nitze, visits the station, December 1963. Note the bureau number on the tail of P2V Neptune, #131410, assigned to VP-5. When a P2V Neptune was donated back to the station by a private individual for use as a static display, it was this same aircraft. (U.S. Navy)

Above: Hangar 1000 with new construction continuing in Segment 5, late 1969. (U.S. Navy)

Left: The first RA-5C Vigilante arrives for rework at the Overhaul and Repair Department, January 1964. (U.S. Navy)

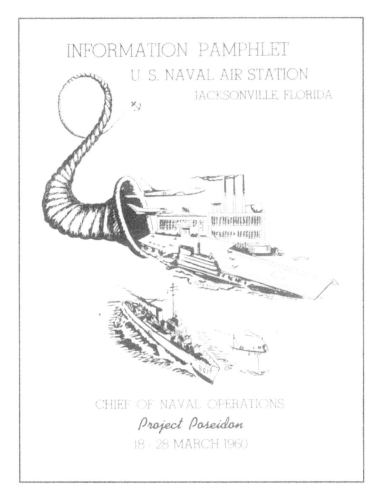

Mrs. John Towers reads the plaque designating the station airfield "John Towers Field" after her husband on October 14, 1960. (U.S. Navy)

Main administration building for the Naval Air Technical Training Center, located on Yorktown Avenue, 1967. (U.S. Navy)

The first A4E Skyhawk awaits rework at the Overhaul and Repair Department in 1966. (U.S. Navy)

NAS Jacksonville contestants greet the Blue Angels prior to performance at the station's 23rd open house anniversary, October 11, 1963. (U.S. Navy)

The first A-7A Corsair (#152657) is inducted to the Naval Air Rework Facility on March 16, 1967. (U.S. Navy)

This aerial shows all of the aircraft assigned to the Naval Air Reserves in 1964. From far left, clockwise: R5D Skymaster, TV-2 T-Bird, S2F Tracker, T34B Mentor, P2V Neptune, A4D Skyhawk, HSS-1N Seabat, SNB-5 Navigator. (U.S. Navy)

Aerial of the main gate taken in 1963. (U.S. Navy)

A TBM-3E Avenger became a static display in front of the NAS Jacksonville administration building, May 1967. The plane was removed a short time later for reasons unknown today. (Photo by William Swisher)

F2H-4 Banshee sits on the ramp in 1961. (U.S. Navy)

T-33B training jet attached to station, March 28, 1967. (U.S. Navy)

An F4H-1F Phantom 11 on the ramp, 1967. (Photo by William Swisher)

This Douglas JD-1 Invader was assigned to VU-10 detachment. It's mission was to tow targets. They were disestablished in July 1964. (U.S. Navy)

P2V Neptune assigned to Patrol Squadron 7. (U.S. Navy)

The station utility plane, a UIIA "Aztec," April 1967. (U.S. Navy)

P2V Neptune assigned to Patrol Squadron 30 on seawall ramp, May 24, 1967. (Photo by William Swisher)

AD-6 Skyraider of VA-35 which was stationed at NAS Jacksonville from October 1958-1965. (U.S. Navy)

Douglas A-1E Skyraider of station-based attack squadron VA-45, May 24, 1964. (Photo by William Swisher)

Skyraider assigned to VA-176, the last to fly the Skyraider when it was phased out April 25, 1968. (Photo by William Swisher)

A3D Skywarrior assigned to Heavy Photographic Squadron VAP-62. (U.S. Navy)

P2V Neptune assigned to VP-7 leaves the station in 1965. (U.S. Navy)

P-3B Orion assigned to VP-30 on April 28, 1967. (Photo by William Swisher)

C-117D assigned to station on May 24, 1967. (William Swisher via Steve Ginter)

C-131F assigned to station May 24, 1967. (William Swisher via Steve Ginter)

US-2A assigned to station May 24, 1967. (William Swisher via Steve Ginter)

Station search and rescue plane, HU-16C Dumbo, May 24, 1967. (William Swisher via Steve Ginter)

VP-49 P-3A Orion, January 22, 1964. (U.S. Navy)

Left: WC-121N Warning Star belonging to Hurricane Hunters, flying to another storm, August 1967. Note Hurricane Dora painted on aircraft. (U.S. Navy Photo #K-64105)

Left: P-3B assigned to VP-45 flies over the City of Jacksonville in 1965. (U.S. Navy)

LCDR John McCain released as a prisoner of war from Vietnam, arrives at the station, March 6, 1973. (U.S. Navy)

Commander Tactical Air Atlantic is established at NAS Jacksonville in a ceremony on May 1, 1973. (U.S. Navy)

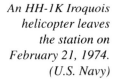

An HH-1K Iroquois helicopter leaves the station on February 21, 1974. (U.S. Navy)

1970

The new $7-million, 400 bed Naval Hospital at NAS Jacksonville was reduced to total darkness on Friday, January 23, 1970, when an electrical fire crippled its lighting system at about 10:00 A.M. Fortunately, only two of the 386 patients were undergoing surgery at the time. One of these patients was immediately transferred to a Jacksonville hospital, while the other operation was completed as aides held battery-operated battle lanterns. The entire hospital continued to operate for two more days, relying upon 670 borrowed battle lanterns for light, until normal power was restored.

The Chief Petty Officers Club opened March 5, with a large number of civilian and military dignitaries in attendance. Also in March, the Naval Air Rework Facility (NARF) started processing the first A7E Corsair II, as it phased out the earlier A7A and A7B versions.

In April, four NAS Jacksonville personnel received citations for having saved a small girl's life back on January 29. On the afternoon of the accident, Mrs. Moore and her eight-year-old daughter, Phyllis Elaine, had been fishing aboard the station at the edge of Lake Casa Linda when a tractor pulling a lawnmower backed into them and knocked them both into the water. A NARF employee, witnessing the incident, jumped into the water and saved the struggling girl with the assistance of three Navy personnel from the Naval Air Technical Training Center.

The first of four new HH1-K helicopters arrived on June 1 to replace the aging H-34s for search and rescue missions. And the station's last C-131 Transport left on June 12 for reassignment to Naples, Italy.

Back in 1948, as NAS Jacksonville became a fleet operational station, the great PBY fleet had departed and seaplane activities had all but ended. Since that time the only remaining seaplanes were two "Dumbo" HU-16 Albatrosses, which remained until July 14, 1970. These last two seaplanes were flown to Arizona to join the mothball fleet. Their departure marked the end

of the station's long-range search and rescue capability.

Captain John R. Kincaid relieved Captain Cutler on July 17, 1970. Captain Kincaid had been credited with shooting down five Japanese aircraft in WWII, and hitting four more.

Also in July, groundbreaking ceremonies were held for the $1 million fuel accessories building at NARF, and for the $2 million domestic and industrial waste treatment facility. The waste treatment facility was designed to handle 2.5 million gallons of waste a day.

A fire in an on-base house claimed the station's first non-aviation related victim on August 25. Mrs. Duncan had been severely burned by the time she was rescued from the burning house and she died two days later. The origin of the blaze was never determined.

The Barnett First National Bank, originally established at NAS Jacksonville back in 1943, held groundbreaking ceremonies for their Yorktown Avenue facility on October 21. Antiquated building destructions commenced in November when 954 and 955 were demolished for a new Navy Exchange parking lot and Buildings 600 and 601 were demolished to make way for the Acey Deucey Club.

On November 1, 1970, Patrol Squadron 62 was established as part of the Naval Reserve Force Squadrons. The squadron initially flew the P2V Neptune patrol aircraft, which had been phased out of all of the stations operational patrol squadrons years earlier. This was only temporary, however, as the P-3 Orions quickly joined the squadron in 1971 and replaced the P2V's.

Finally the Fleet Air Jacksonville Band Unit 191 was disestablished on December 16. The band had been a long time feature at almost every significant event at NAS Jacksonville since 1948 and everyone was sorry to see it go.

1971

In January 1971, a functional name change occurred as the Aircraft Maintenance Department became the Aircraft Intermediate Maintenance Department (AIMD). The Aircraft Maintenance Department had been established in 1959 at NAS Jacksonville

when the Fleet Aircraft Service Squadrons (FASRons) were phased out. The current AIMD is part of the Three "M" (3M) system, encompassing: First, or Organizational, Level Maintenance, performed by operating squadrons who provide organizational maintenance and consisting primarily of the removal and installation of components and routine field repairs; Second, or Intermediate, Level Maintenance, which is performed by AIMD and includes limited repair of engines and overhaul of certain accessories and components, and; Third, or Depot, Level Maintenance performed by the Naval Aviation Depot, which does complete reworks and serves as industrial backup for the AIMD. The comprehensive 3M system remains a cooperative effort in which each level is essential to the combat readiness of the fleet operating forces.

On January 15, Construction Battalion Unit 410 (CBU-410) was organized in ceremonies in front of Building 1. Their commissioning was part of a Navy-wide effort to develop skilled craftsmen for rapid use in required operations.

Also in January, the "Hurricane Hunters" received their first P-3A Orion, and commenced phasing out the NC121N Super Constellation.

On April 27, the new Barnett Bank facility opened and in May, Marine Private First Class Richard Rice became the one millionth student to graduate from the Naval Air Technical Training Center. After brief ceremonies at NAS Jacksonville, Mr. Rice was flown to Memphis for further ceremonies.

On July 2, VP-56 became the first squadron to transfer from NAS Patuxent River, Maryland. VP-49 and VP-24 followed in 1972.

The air station took the brunt of a few "jokes" on July 5, when they fired a 21-gun salute to celebrate the 4th of July one day late. When asked why the ceremony was not conducted on the appropriate day, the NAS duty officer simply responded "We didn't get around to doing it Sunday, so we did it today."

July 11 was truly an eventful day for NAS Jacksonville, commencing with an "Open House" and air show, featuring the Blue Angels, the Chuting Stars, a helicopter rescue demonstration and other activities which attracted more than 80,000 people to the station. Then, as the last of the Sunday air show traffic jams were clearing, a devastating fire broke out in Building 551 and in an adjoining building in the Naval Air Technical Training Center area. Although the exact cause of the estimated $1 million fire was undetermined, lightning was strongly suspected since a severe electrical and thunderstorm was

in progress at the time the fire started. Several explosions, including three large ones in the southeast corner of the complex, sent firemen scurrying for cover. Because extremely low water pressure in the base hampered the firemen's efforts, the two buildings were totally destroyed and another building was

extensively damaged. Twelve firemen were hospitalized for smoke inhalation and minor injuries. Fortunately, no one was seriously hurt.

The first C5A Galaxy (US Air Force plane) landed at NAS Jacksonville on August 20. It was the largest aircraft to ever land on the station's runways.

On October 1, the new $650,000 Petty Officers Mess, Hancock Hall, opened. The facility was named after the late Hospital Corpsman Second Class E. Scott Hancock, a Robert E. Lee High School graduate, who was killed while assisting wounded Marines on February 24, 1969, just 11 days after entering Vietnam.

For his valor, Hancock was posthumously awarded the nation's second highest medal, the Navy Cross.

On December 9, the first F-14 Tomcat landed at NAS Jacksonville.

1972

Wednesday, January 26, 1972 saw the famous VP-49 "Woodpeckers" arrive at their new homebase at NAS Jacksonville. When asked by Admiral Geis why they were called "Woodpeckers," a squadron member explained "It takes a hard nose -like a woodpecker's beak - to get tough jobs done in antisubmarine warfare (ASW)."

Environmental concerns surfaced on March 2 when 1000 gallons of JP-5 (jet fuel) were accidentally spilled into the St. John's River. As a result, the Naval Air Rework Facility (NARF) installed automatic shut-off valves on their tanks to preclude future tank overflows.

VP-30 transferred a detachment

back to NAS Jacksonville on May 1. The main squadron would not relocate for another two years. Later in the month, soul singer Lou Rawls gave two memorable performances at the CPO Club and at Hancock Hall on May 24.

Captain William G. Sizemore assumed command of NAS Jacksonville on July 1. Captain Sizemore was no stranger to NAS Jacksonville, having received his gold naval aviation wings after flight training at the station in 1948. On July 10, the Hurricane Hunters retired their last NC-121 Super Constellation aircraft. This plane had been a familiar sight over the skies of Jacksonville for almost 20 years.

On October 1, Lt. Mark Graley, the first POW to be released from Vietnam, entered the Naval Hospital Jacksonville for a thorough examination and debriefing.

Finally on November 30, the 983 officers, enlisted men and dependents of VP-24 arrived at their new homebase.

1973

On January 12, 1973, Commander Frank Woodlief made history by becoming the first Naval Flight Officer (non-aviator) in the Navy to command an antisubmarine warfare patrol squadron when he took command of the oldest squadron at NAS Jacksonville, the "Mad Foxes" of Patrol Squadron 5. Commander Woodlief was one of the Navy's new breed of electronics and weapons delivery systems wizards whose knowledge and experience were opening up many billets that had been previously reserved for pilots only.

NAS Quonset Point, Rhode Island's loss became NAS Jacksonville's gain on April 16, when defense cuts and functional requirements forced the closing of NAS Quonset Point and caused the helicopter squadrons based there to be reassigned to NAS Jacksonville. Along with these squadrons came Helicopter Antisubmarine Wing ONE, which controlled the helicopter operations.

Fleet Air Jacksonville, established at NAS Jacksonville in 1948, was decommissioned on June 29, 1973. The six VP antisubmarine squadrons of Fleet Air Wing Eleven, comprising 2100 men and 54 P-3C Orion aircraft, were shifted under the new command title of Commander Patrol Wing Eleven, as part of a Navy-wide project to reduce land-based staffs.

August 1973 saw the Naval Air

Rework Facility complete their last A-4 "Skyhawk" rework. When the program ended, they had reworked over 1200 of these aircraft.

The first of the newly arriving helicopter squadrons, the "Shamrocks" of Helicopter Squadron (HS) 7, landed on September 27, followed by the "Sea Horses" of HS-1 on October 13, the "Tridents" of HS-3 and the "Dragon Slayers" of HS-11 on October 17, and the "Red Lions" of HS-15 on November 5. The new squadrons would provide search and rescue as well as antisubmarine warfare capabilities. The Helicopter Antisubmarine Wing One, who controlled the helicopter squadrons, moved on December 15. On October 25, 1973, the Naval Air rework Facility accepted the first S-2F Tracker for rework.

The Naval Air Technical Training Center (NATTC) was officially closed for the second time on November 1, 1973, although the final 460 students did not leave until their graduation the following February. The closing occurred so that once again the NATTC could be consolidated with the NATTC at Memphis, Tennessee. On the same day, ground was broken for a new $3 million operational and maintenance training facility for the P-3 "Orion." This facility, with training spaces for weapons systems trainers, flight simulators and maintenance trainers is still used extensively today, as plans for another addition is reaching the final stages. November 3-4 saw another air show at NAS Jacksonville. This show was previously scheduled for April 14 and 15, 1973, but was postponed and rescheduled just five days prior to the start of the air show. "The Shirelles" closed out the year with performances at Hancock Hall on November 16 and by Glenn Miller and his orchestra at the CPO Club on December 27.

1974

The "Nightdippers" of HS-5 continued the helicopter squadron transfers begun in 1973 when they arrived on February 1, 1974. Eventually, Helicopter Antisubmarine Wing ONE would be composed of six operational squadrons and one training squadron, HS-1. The squadrons arrived flying the Sikorsky SH-3 "Sea King" but all fly the SH-60 "Seahawk" today. On February 21, with the graduation of Aviation Electricians Mate Class #411, the Naval Air Technical Training Center transferred to NAS Memphis. This would be the second time the schools would be closed and transferred away from the station.

In June, the newly created Navy Campus For Achievement program came to NAS Jacksonville. The Navy Campus Network, as it is now known, consists of a world-wide system of professional education specialists whose mission is to establish, promote and manage all on base civilian education programs.

On July 12, RADM Geis had an unusual change of command. He had been performing two different functions. His first function, as the Admiral in charge of the Tactical Wings Atlantic, was given to RADM John Dixon, who moved that command to NAS Oceana. Then RADM Green assumed command of Commander, Sea Based ASW Wings Atlantic, the senior command on NAS Jacksonville. This also marked the official end to Commander, Fleet Air Jacksonville.

September 20 saw another unusal event as 10,000 visitors came to the NAS Jacksonville seawall along the St John's River to view the Admiral's Cup HydroPlane (boat) races. Miss Budweiser won the race.

November 16-17 again saw the Blue Angels perform at another air show event. This would also be the last air show held at the station for 16 years. Prior to the air show on 16 November, the Florida Society of Colonial Dames, along with base Commanding Officer Captain Bernstein, unveiled the Mulberry Grove Plantation marker. This marker replaced the marker that had been at this spot with the wrong historical information since 1940.

1975

In January 1975, groundbreaking ceremonies were held for the JAX Navy Federal Credit Union's new building. This modern facility, built at a cost of $220,000, was twice as large as the one it replaced. As of the date of groundbreaking, the credit union had 44,000 members with assets in excess of $60 million.

In March, the "Pro's Nest," Patrol Squadron THIRTY (VP-30), began returning to NAS Jacksonville from NAS Patuxent River, Maryland. The move was not completed until August. The squadron was one of two replacement patrol squadrons in the Navy, which trained pilots, flight

officers, aircrewmen and maintenance personnel in the operation and upkeep of the P-3 "Orion."

In April, NARF started rework on the new TF-34 jet engine, used on the Cecil Field based S-3A "Viking." The "Viking" was the Navy's newest antisubmarine warfare aircraft. Also in April, NARF rolled in the first P-3 "Orion" for rework. Since then, over 300 P-3 "Orions" have been reworked at NARF, with the average rework period lasting 16 weeks. In 1975, NARF employed

2,800 civilians and had an annual payroll of $45 million.

On April 30, the "Hurricane Hunters" of VW-4 were decommissioned due the advent of new satellite technology. (The formal ceremony was held on April 23.) From the time the Navy first started to get serious about hurricane tracking and prediction in 1943, until the "Hurricane Hunters" were decommissioned, the death rate per storm had dropped from 400 lives to 4, a truly remarkable achievement.

On May 4, 1975, the first of the year's visiting dignitaries arrived. King Hussein of Jordan and Shah Paheui of Iran landed and were escorted to Cecil Field where they saw an impressive display of jet aircraft. NAS Jacksonville was again in the world spotlight November 1 through 4, as President Anwar Sadat of Egypt, President Gerald Ford and Secretary of State Henry Kissinger, arrived at the station. After a short stay at NAS Jacksonville, the three took Navy boats to Epping Forest, for talks on the Middle East. The actual significance and results of their meetings remains a mystery to all but the three participants.

1976

The bicentennial year of 1976 began in a spectacular manner when a Marine AV-8 Harrier jet, in flight from Dobbins Air Force Base near Atlanta to NAS Jacksonville, crashed between the airstrip and a taxiway and exploded. The plane's pilot had almost managed to bring his vertical take-off and landing aircraft to the hover position when the plane unexpectedly made a sharp left turn. Forced to eject at an altitude of about 100 feet, the pilot escaped with minor injuries. The aircraft was totally destroyed. Also in January, the first complete degree program came to NAS Jacksonville when Southern Illinois

University (SIU), initiated on-base college classes. The SIU B.S. in Vocational Education degree program was highly innovative in that it gave military personnel extensive credit for their service schools and work experience. Over the years, the SIU program has grown to become the most popular B.S. degree program offered at NAS Jacksonville.

The first P-3C "Orion" flight simulator commenced operation at FASO in April and the HS-9 "Sea Griffins" were re-established under Helicopter Antisubmarine Wing One at NAS Jacksonville on June 4, 1976.

CAPT Wayne D. Bodensteiner became the 23rd Commanding Officer of Naval Air Station, Jacksonville, on Wednesday, July 14. He relieved CAPT Karl J. Bernstein, who was promoted to the rank of Rear Admiral. One week before the change of command, groundbreaking ceremonies for the new $1.9 million Armed Forces Reserve Center were held. The new 6,100 square foot, two story building on U.S. 17 was located on the same site where German POWs had been served dinner during WWII. Additional groundbreaking ceremonies, this time for the new $472,478 aircraft fire rescue station, were held Thursday, July 22.

NARF, where the first A-7 Corsair had been reworked back in September 1967, reworked the exact same aircraft again in August 1976. As of August 1989, over 900 A-7 Corsairs had been reworked by NARF personnel. In September, a Memorandum of Understanding (MOU) was signed with Central Michigan University (CMU) to bring their Master of Science in Administration (MSA) degree program on board. September 28 saw Secretary of the Defense Donald Rumsfeld visit the station.

1977

Actor Mike Farrell (B.J. Honeycut on the M.A.S.H television series), visited personnel at the Naval Hospital in March 1977. Two weeks later, the Jacksonville Operating Area Coordination Center (JOACC) at NAS Jacksonville became the Fleet Area Control and Surveillance Facility (FACSFAC). FACSFAC's mission is to schedule and control offshore Navy fleet operating areas,

military special use airspace, land target and electronic warfare operations. In more basic terms, as Navy aircraft leave NAS Jacksonville, NAS Cecil Field or Mayport NS, the NAS Jacksonville air traffic controllers initially control them before "handing them off" to departure control at the Jacksonville International Airport. After reaching a certain altitude and distance, they are picked up by the Federal Aviation Administration Control Center at Hilliard. Hilliard then guides the Navy aircraft to military "warning zones," where FACSFAC personnel take over. FACSFAC also helps to control the 350 plus civilian aircraft a day that fly near the military warning zones.

April 11 saw the new Consolidated Package Store open at the station. This was one of the more unique locations to place a store where sailors could buy alcoholic beverages. It was located right across the street from the Alcohol Rehabilitation Center! It was the subject of more than one joke around the station for years.

In June 1977, the Enlisted Mens' Club (Pirate's Cove), was co-located with the Petty Officers' Club at Hancock Hall. The combined clubs were now called the "Enlisted Mens' Mess (Open)" (EMO).

On September 30, Helicopter Combat Support Squadron Two (HC-2), the oldest helicopter squadron in the U.S. Navy, was disestablished. The squadron had recorded over 2,000 rescues over the years. September also saw 19 military and civilian students receive their B.S. degree in Education from Southern Illinois University in the first formal college commencement to be held on base. Since then, over 400 students have graduated from this program. A month later, the first group of Central Michigan University students completed their MSA degrees.

On November 1, Fleet Logistics Support Squadron Fifty-Eight (VR-58) was established at NAS Jacksonville. VA-103, a reserve jet squadron, was relocated to NAS Cecil Field to make space for the new squadron. VR-58 flies the C-9B jet transport and its basic mission is to train reserve aircrew and maintenance personnel in transport operations. The new squadron's personnel budget brought another $1 million annually to the Jacksonville economy. This fact did not go unnoticed by the community, and when first C9-B jet was received by VR-58 in early 1978, Jacksonville Mayor Hans Tanzler christened it the "City of Jacksonville."

December 1 saw reserve squadron VA-203 reassigned from NAS Jacksonville to NAS Cecil Field. The squadron was flying the A-7 Corsair at that time. With this move, there were no attack/fighter jet aircraft assigned to any squadron at the station for the first time since the station hosted the Navy's first Atlantic Coast jet squadron in 1949.

1978

Groundbreaking ceremonies were held for the new $1.3 million Naval Regional Medical Center Branch Clinic and Dental Center in January. The new Disease Vector Ecology and Control Center (DVECC) building was dedicated in January, as well. Today, as one of the most advanced centers of its kind in the Navy, DVECC's mission is to conduct research into shipboard pest control problems. DVECC personnel also conduct training for ship and shore personnel on how to control pests. Aerial pesticide dispersal was a special field at the center's research and testing service.

On June 27, 1978, the Jacksonville City Council passed an Air Installation Compatible Use Zone ordinance that was a landmark piece of legislation. As civilization encroached and eventually surrounded NAS Jacksonville, accidents involving Navy and civilian aircraft, plus the problem of increased jet noise, had become major concerns to the civilian community located near the station. NAS Jacksonville officials were already "sensitized" to the problems and had instituted a number of restrictions, including limited aircraft flying areas near civilian areas, restricting aircraft engine run-ups between the hours of 10 P.M. and 7 A.M., transferring Jet Attack Squadron 203 to Cecil Field and prohibiting jet practice landings at NAS Jacksonville. While these measures helped to alleviate the current problem, the legislation passed by the City Council would control future land development near NAS Jacksonville, NAS Cecil Field and NS Mayport. Accident zones around the stations were divided into three classes: A, B and C. Noise zones were also placed in three classes. Accident zones marked "A" offered the most serious threats, and the Navy attempted to purchase all "A" zones. Included in the bill was a "truth-in-sales" provision that requires a seller of land in the affected zones to notify persons buying or renting. Both Navy officials and the City of Jacksonville leaders were pleased with the ordinance, which remains in effect today.

The Armed Forces Reserve Center's new $1.9 million facility, which included an indoor firing range, was dedicated on July 15.

Captain Joseph M. Putrell relieved Captain Wayne D. Bodensteiner on August 8. One week later, the Naval Regional Data Automation Center (NARDAC) was established. NARDAC's mission was to assist in the design, development and maintenance of Navy computers (automated systems). AIMD also set a production record in August, repairing 59 P-3C Orion engines in 57 days while simultaneously undergoing a major renovation.

The Jacksonville Navy Federal Credit Union reached $100 million in assets on September 8, and in November, one of the most advanced jet engine test facilities in the Navy opened at NARF. The new facility was renamed the Richard J. Kemen Standard Navy Test Cell, in honor of the University of Florida mechanical engineering graduate who had been a NARF engineer since 1962, and

whose interest in jet engine testing led to the development of the test facility, which now bears his name. This facility leads the Navy in gas turbine engine testing with facilities for testing all turbojet/turbofan jet engines.

1979

The Personnel Support Activity was established January 29, in an effort to streamline procedures, decrease errors, and provide better service to military personnel. A reduction in errors with military pay was of concern to most military personnel. To handle civilian pay, the Regional Accounting and Disbursing Center (RAADC) was also established at this time. RAADC handled civilian pay for all shore activities in Florida in addition to the civilian personnel at NAS Jacksonville.

On March 21, the station Search and Rescue (SAR) helicopter crashed in the Atlantic, killing three. This was the last crash of an aircraft-assigned directly to NAS Jacksonville proper (not a squadron aircraft) through the printing of this book in 2000. This particular SAR helo was on a training mission. The lower back section of this helicopter, composed of a door, was capable of being lowered down for loading and unloading. This was apparently done during flight and while hovering low in the Atlantic off the coast of Jacksonville.

The back section of this helicopter with this door extended down got too low and entered the Atlantic. The aircraft started filling up with water. The pilot and co-pilot were able to escape through the front. Three of the aircrew in the back were not so lucky, however.

The base switchboard, installed in 1940, was finally modernized on May 15. On July 24, the world's largest aircraft, a C-5A "Galaxy," again made a rare visit to the station. In October, the station Security Department increased its capabilities by joining on the Computerized On-Line Police System (COPS). This system allows instant access to vehicle registration information. Finally on November 15, as the decade of the 70s was coming to an end, the Naval Regional Medical Center's Health Care Facility and Dental Center was dedicated.

CHIEF PETTY OFFICERS MESS (OPEN)

NAVAL AIR STATION
JACKSONVILLE, FLORIDA

OFFICIAL OPENING

2000

5 MARCH 1970

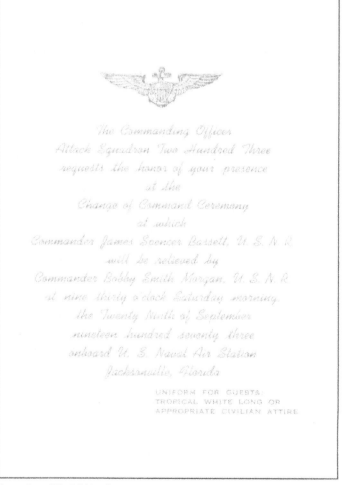

Squadron VA-203's change of command ceremony invitation. This was the last attack squadron assigned to the station.

Squadron Disestablishments of the 1970's

VW-4 Hurricane Hunters
Disestablished April 30, 1975

Navy airborne hurricane reconnaissance actually began in 1943. PBM Mariners starting investigating tropical storms. VPB-114 became the Navy's first hurricane squadron in 1944, based in Boca Chica, Florida, and partly in San Juan, Puerto Rico. This squadron was designated Weather Squadron Three (VPW-3) in late 1945. This eventually lead to the formation of Patrol Squadron 23 at NAS Miami in 1949. This squadron hunted hurricanes with Privateer patrol bombers. In 1952, Weather Squadron Two (VJ-2) was established at NAS Miami, which replaced VP-23. Shortly after formation, the squadron moved to NAS Jacksonville and received new P2V-3W Neptunes during the 1953 hurricane season. On December 15, 1953, VJ-2 was redesignated Airborne Early Warning Squadron Four (VW-4).

VW-4 took delivery of its first WV-1 Constellation in late 1954. This larger, four-engine plane augmented the P2V Neptunes but didn't begin actual storm penetrations until 1955 when WV-1 had its first confrontation with Hurricane Diane. The WV-1's were replaced by WV-3 Super Constellations in early 1955. The P2V-3W's were also replaced by P2V-5F's, which had the advantage of jet engines.

The squadron lost its only aircraft in 1955 while penetrating the eye of Hurricane Janet. No trace of the aircraft or its eleven-crew members, along with two Canadian newsmen, was ever found.

By 1958, all of the Neptunes were retired and the squadron was exclusively flying the Super Constellation. In September 1958, another milestone was reached when a Super Constellation flew into two different hurricanes, Helene and Ilsa, on the same day.

On August 16, 1960, the homeport of the Hurricane Hunters was changed to Naval Station Roosevelt Roads, Puerto Rico. The last of the dependents were moved by December 13, 1960. The squadron returned to NAS Jacksonville, however on January 15, 1965.

On March 1, 1967, the unit's title was changed to more accurately reflect their mission - Weather Reconnaissance Squadron Four.

WP-3A's replaced the Super Constellations in 1971 as the first Orion arrived on January 29, and the last Super Constellation left the squadron on July 10. The Super Constellations had flown over 70,000 accident free hours.

In June 1974, LT(JG) Judy Neuffer joined the squadron, as the second woman designated a Naval Aviator. In September she became the first woman to pilot a plane into a hurricane, as she piloted an Orion into the eye of Hurricane Carmen.

On April 30, 1975, VW-4 was disestablished.

HC-2 Fleet Angels
Disestablished September 30, 1977

Helicopter Squadron 2 was established as Helicopter Utility Squadron 2 at NAS Lakehurst on April 1, 1948. Along with HU-1, these squadrons formed the first operational helicopter squadrons in the Navy. The squadron's original aircraft was composed of the Sikorsky HO3S and Bell HTL. Supported by some 300 officers and men, the mission was to train helicopter pilots, all of whom were veterans of some previous military fixed wing experience. Captain C.C. Marcy became the first Commanding Officer.

After just one year of service, the squadron was flying plane guard, transfer of personnel, U.S. and guard mail service, radar calibration, photographic flights, and torpedo tracking and reconnaissance. By the middle of the 1950's almost 60 percent of the squadron's total flight time was logged at sea. Detachment One was permanently homeported at Norfolk with Search and Rescue Detachments at NAS Atlantic City, New Jersey, NAS Jacksonville and NAS Key West, Florida. The first HUP-1 helicopter was delivered to the squadron on January 9, 1951. In late 1950, HU-2 was called upon to deploy several detachments to the west coast, with only 72 hours notice, to operate with the Pacific Fleet during the Korean Conflict, and subsequently saw some action. Commander, Naval Air Force Atlantic, authorized night flying on January 29, 1952. By the end of 1952 HU-2 had 52 detachments serving on several classes of ships from the USS Missouri to the USS Franklin D. Roosevelt. Total rescues numbered 290 and the squadron now had 435 enlisted men and 106 officers. By the end of 1955 the squadron had recorded 374 rescues. In 1957, HU-2 became the first operational squadron to receive the new Bell HTL-5 helicopter. Floats were included in this helicopter to allow for landing at sea. By 1958, the squadron had an impressive aircraft inventory number with 58 assorted airframes. The HUK-1 helicopter was also introduced at that time.

The 1960's saw detachments involved with the Mercury space program as Detachment 36 transferred LTCOL John H. Glenn from the USS Noa to the USS Randolph after recovery. During the Cuban Missile Crisis of October and November 1962 HU-2 Detachment 65 aboard the USS Enterprise and Detachment 62 aboard the USS Independence flew in support of the quarantine. Both detachments flew 864 hours during the 49-day period. Cargo transfers amounted to 45,000 pounds and 3 rescues were recorded. 1962 also saw the squadron start flying the Kaman UH-2A Sea Sprite. In July 1965, HU-2 was redesignated Helicopter Combat Support Squadron Two, HC-2. The new squadron insignia was also approved. The year 1965 saw the squadron already recording rescues from Vietnam. By 1967, the squadron had rescued more than 1,500 personnel.

On February 15, 1971 the squadron took delivery of its first SH-3G Sea King helicopter. The Sea King offered extended range, longer time on station and a more sophisticated electronics system than the squadron had previously experienced. In 1973 the "Fleet Angels" began another transition period. Unlike similar operations in the past, this affair was not to involve flying a new helicopter. Rather it involved moving from NAS Lakehurst, New Jersey, HC-2's home of 25 years, to NAS Jacksonville. Commander Mike Marriott's farewell message to the squadron contained reference to the move: *"It was not easy leaving that friendly, small Naval Air Station. The very things we complained about the most, the smallness, the remoteness and the humble but personal services of the dispensary, exchange and commissary are the same general problem areas we find at Jacksonville. The hospital is big, but impersonal. The commissary is large, but crowded. The high perch we sat on at Lakehurst by virtue of our established seniority and the ability to do our job second to none is not readily available here in Jacksonville - But is coming."*

The following year HC-2 was placed under the operational control of Helicopter Antisubmarine Wing One of Jacksonville and the rumors of disestablishment were starting to circulate. With a growing number of HS squadrons on the east coast, the mission that HC-2 had performed exclusively for almost 30 years was drying up. By 1974 the squadron had only three active detachments.

On a dark night in January 1977, about 70 nautical miles south of Palma, the last rescue was recorded. The crew of a burning marine Phantom from VMFA-133 was successfully rescued by two Detachment 2 aircraft. The final total for rescues for this squadron was 2,318. On September 30, 1977, HC-2 was finally disestablished at NAS Jacksonville.

VA-203 transferred from NAS Jacksonville to NAS Cecil Field on December 1, 1977.

After an apparent tail rotor problem, the floatation bags were deployed and this aircraft floated in the St. Johns River. Tow boats were dispatched to bring it to the pier, where a crane lifted it for repairs back at the hangar. (U.S. Navy)

LTJG Everett, LCDR Raebel, CDR Palfer and LCDR Souder, all former POWs, "Meet the Press" at a news conference held April 13, 1973 at the Naval Hospital Jacksonville. (U.S. Navy)

The last WC-121N Super Constellation departs the station in a ceremony on July 24, 1972. (U.S. Navy)

An aerial view of the USAF 679th Radar Squadron radar tracking installation taken July 16, 1972. This tower was demolished in September 1981. (U.S. Navy)

VP-62 formed initially flying these P2V-7 Neptunes in 1977. (Robert Kling)

A fire truck gets into position to fight this fire which consumes the east cooling tower for the Naval Air Rework Facility's jet engine test cell. The tower, which was in final construction, was completely destroyed and had to be totally rebuilt. (U.S. Navy)

Above: A view of the Naval Air Technical Training Center fire taken July 11, 1971. The fire started as the last of the station's earlier air show crowd was leaving. (U.S. Navy)

Right: Three Corsairs, a YA7N, F4U, and A7E, sit in front of Hangar 1000 on November 3, 1973. (Bill Sides)

The last of the observation towers used when seaplanes were based at the station is removed from Hangar 140 in 1970. (U.S. Navy)

A pilot goes down the original "Dilbert Dunker" in 1975. (U.S. Navy)

LT Judith Ann Neuffer, the second female pilot in the Navy to earn her wings, stands next to a WP-3A Orion assigned to the Hurricane Hunters on September 16, 1974. (U.S. Navy)

VA-203, A4L Skyhawk in Hangar 113, September 1973. VA-203 phased out their A-4s at NAS Jacksonville on March 8, 1974. (Bill Sides)

Station-based US-2B in 1973. (Bill Sides)

A7A Corsair assigned to VA-203. This squadron left the station in June 1978 for NAS Cecil Field. (Bill Sides)

P-3C Orion assigned to the Hurricane Hunters. (U.S. Navy)

A4L Skyhawk assigned to VMA-142 on the ramp. (Bill Sides)

P3C Orion assigned to VP-5, November 1973. (Bill Sides)

T28B assigned to station for pilot training on ramp, June 24, 1973. (Bill Sides)

P-3C Orion assigned to VP-24, March 17, 1976. (U.S. Navy)

P-3C Orion assigned to VP-45, November 1973. (Bill Sides)

HS-3 Sea King protect the aircraft carrier during independent steaming. (U.S. Navy)

P-3C Orion assigned to VP-49, May 1973. (Bill Sides)

SH-3A approaches for a landing on the flight deck of USS Dwight D. Eisenhower (CVN-69). The squadron is HS-5. (U.S. Navy)

P-3C Orion assigned to VP-56. Gen. Alexander M. Haig Jr., Supreme Alied Commander, Europe, was among the passengers when this photo was taken April 15, 1977. (U.S. Navy)

HS-9 Sea King helicopter operates with a Brazilian SH-3. (U.S Navy)

P3A assigned to VP-62, February 1972. (Bill Sides)

HS-75 operates from the deck of USS Dwight D. Eisenhower (CVN-69), May 1984. (U.S. Navy)

SH-3 Sea King assigned to the Fleet Angels of HC-2. (Robert Kling)

Above, HS-11 Sea King takes off from AUTEC, Andros Island, Bahamas. Note torpedo on starboard side. (U.S. Navy)

Right: P-3C Orion (update II) assigned to VP-30 flies over the station on December 8, 1977. (U.S. Navy)

Below: H-46 Sea Knight flies over downtown Jacksonville, March 1971. Three of these helicopters were assigned to the station for search and rescue. (U.S. Navy)

1980

The year 1980 saw a Presidential visit to the station when President Carter made a brief appearance at Jacksonville. The Naval Air Reserve Unit and Readiness Command's present building was nearing final completion and the Flying Club, located for years at Herlong Field, transferred back to the station. Captain William C. Christenson became the 25th Commanding Officer of NAS Jacksonville on August 28.

One Sunday evening in November, a small Cessna V-206 Super Skywagon landed undetected at the station. When discovered by a maintenance person the next morning, 542 pounds of Colombian marijuana valued at $250,000 was found on board. An investigation of this unprecedented incident revealed that the plane had landed and taxied near the flying club where the pilot then siphoned 174 gallons of fuel from four-single engine planes. When he tried to take off again, he found the plane would not start. Undaunted, he next tried to jump-start the Cessna with a battery taken from a "yellow-gear" tractor. When the plane still refused to start (due to a broken alternator belt), the pilot evidently just walked off the station, abandoning both the plane and drugs. He was never found.

VR-58 dedicated their second C-9B Sky Train, "City of Orange Park," in December.

1981

A suspicious fire in Building 906 in April, totally destroyed the structure where the Naval Investigative Service was located. The fire was so intense it cracked the windows on the nearby Naval Air Reserve Unit building, which was to be dedicated in only four days. Two Navy enlisted females who were trying to destroy evidence in an on-going NIS investigation apparently started the fire. Through an intensive investigation by NIS, the females were charged and convicted of arson at a general court-martial. Many other NIS investigations were severely hampered by the fire.

In May, the first Armed Forces Day-Scout World was held on base. A full day's program of Armed Forces and Scouting exhibits were featured and a hot-air balloon race concluded the day's activities. About 4,000 Scouts and their families camped out on base and some 6,000 visitors from the local area came out for the event.

In June, the station's motto, "Service to the Fleet," was put to the test in the wake of the aircraft accident on board the USS Nimitz on May 27. The Nimitz had been operating 60 miles east of Jacksonville when an aircraft crash killed 14 sailors and injured 48 others. Six helicopters from the air station participated in search and rescue operations. All the injured personnel were brought to Naval Hospital Jacksonville for treatment. NARF A-7 program planners were on board the USS Nimitz the next morning to evaluate the eight damaged A-7 Corsairs. Two of the planes were damaged beyond repair, but the other aircraft were trucked from Norfolk, Virginia to NAS Jacksonville, repaired at NARF, and returned to the fleet.

In September, the Air Force Radar Squadron Tower was being removed and memorial services were held for HS-3 Commanding Officer Commander Paul Nelson. Commander Nelson, a 23-year Navy veteran, was one of three Navy men listed as missing in the September 13 crash of a Sea King helicopter from the aircraft carrier USS Forrestal. The three other crewmembers of the six-man crew had survived the crash.

1982

Captain Roger Rich, Jr. assumed command of NAS Jacksonville on July 12, 1982. It was only a short trip from the first floor of Building One where he served on the Staff of Rear Admiral Fred Johnston, Jr., Commander of Sea Based ASW Wings Atlantic, to his new office on the second floor of the same building. Later in July, Captain Rich helped to break ground for the new Child Care Center.

The NAS Supply Department was reformed as the Naval Supply Center (NSC) on October 1. The new Supply Center, which evolved from the consolidation of three local supply activities, became the Navy supply point for the Southeastern United States, the Caribbean, the Gulf of Mexico, the Panama Canal Zone, Guantanamo Bay and Puerto Rico. The NSC stocks over $6 billion in supplies today.

Also in October, a new helicopter training facility was dedicated and named the Paul L. Nelson Helicopter Training Facility in memorial to CDR Nelson, former HS-3 Commanding Officer, who was killed the year before. Housed in the facility are two primary helicopter trainers, one for anti-submarine warfare tactical team training and one for weapons system training.

1983

As Cecil Field was gearing up for the arrival of the first F/A-18 Hornet in April 1983, NAS Jacksonville was the scene of the second worst airplane crash in the station's history. Of the 14 personnel on board their home bound C-131 flight to Guantanamo Bay, Cuba, only one survived the crash. The disaster occurred shortly after takeoff from NAS Jacksonville around noon on Saturday, April 30. The plane had just crossed the St. Johns River, heading east over San Jose Boulevard, when the pilot radioed that his left engine was on fire and he was returning to the base. One minute later, debris from the ill-fated aircraft struck a car on Old Kings Road. Then, as the aircraft was over the St. Johns River about 1/4 mile from the start of the runway and approximately 200 feet up, the left wing separated from the aircraft. When the fuselage hit the water, the plane exploded killing 13 of those on board. The sole survivor, AT2 Melissa Kelly, suddenly finding herself in the St. Johns River, grabbed onto the first floating object she could reach: her own suitcase. The demolished aircraft was eventually raised and sold for scrap.

Life at the station had returned to normal as Armed Forces Day and Scout World approached again in May. The annual event had grown remarkably from 6,000 visitors to 120,000 in just three short years.

In July, Geraldo Rivera was on station shooting footage for an upcoming ABC 20/20 report on the government's drug interdiction program in Florida.

In November, Rear Admiral Allan F. Paulson, Commander of Sea Based ASW Wings' Atlantic, and his wife were critically injured in an automobile mishap. As a result of this accident, in which seat belts were credited with saving the Admiral's life, a stringent seat belt policy was initiated at NAS Jacksonville. The same policy remains in effect today.

In December, the Telecommunications Department, part of NAS Jacksonville since its inception, was absorbed into the newly formed Naval Telecommunications Command.

1984

In February, the Naval Investigative Service was again burned out of its home, which was located in Building 451. Also incinerated in the fire were the offices of the Naval Absentee Collection Unit, some Navy Regional Data Automation equipment and some NARF computer equipment. Building 451 was totally destroyed, along with an estimated $1 million in equipment. Fortunately, the building had already been scheduled for demolition the next year. This fire, as determined by the Fire Marshal, was due to a faulty boiler.

In March, both the new $1.3 million Navy Lodge and the new $1.2 million Child Care Center, opened. The center was designed to accommodate up to 200 children. The fourth annual Armed Forces Day Scout World attracted an estimated 126,000 in May.

On July 12, Captain Roger Rich relinquished command to Captain Lynn C. Kehrli. After the ceremony, Admiral A. C. Paulson promoted Captain Rich to the rank of Commodore. An elated Rich stated "I never graduated from the Naval Academy, and I don't mind that, but what I do mind is that I never got to throw my hat and I'm going to do that now." With that, he glee-fully tossed his captain's hat into the audience! Commodore Rich's new assignment was as commander of the Pacific Fleet Anti-submarine Warfare Wings at North Island Naval Air Station in San Diego, California.

In December, the wage grade employees of NARF voted to bring in the International Brotherhood of Teamsters to represent them and the Teamsters Union has represented these employees ever since.

1985

January shivered in as one of the coldest on record at NAS Jacksonville. As the temperature dropped to 7 degrees, many minor vehicle problems occurred but the major headaches were caused by the water pipes. Numerous fire sprinkler pipes froze and burst, spraying buildings with water and causing extensive interior damage to some. This was especially true in Hangar 123 where pipes burst spraying helicopters, some open for repair, with water.

February 12 saw another fatality of a worker on station. A worker at the Navy Exchange warehouse was moving pallets with a Prime-mover, a device you can stand in and drive around to move pallets. As he backed up the machine, the first level of shelving caught him in the back, and pushed him into the front of his machine control panel, crushing him. An investigation was done and recommendations made to prevent this from ever occurring again.

On March 4, two men were killed when their twin engine Piper Apache aircraft crashed in a dense fog on the NAS Jacksonville runway. The pilot had established radio communication with the NAS Jacksonville control tower and requested to be "talked down." He was also low on fuel. The investigation revealed that the aircraft was loaded with marijuana and that one of the deceased had $4,000 in currency in his possession. A pair of night vision goggles was found with the wreckage. March 15-16 saw Admiral James D. Watkins, Chief of Naval operations visiting the station.

In early April, the McDonald's restaurant opened on station. An interesting story surrounds the location of the restaurant. McDonalds came in with their standard restaurant design and wanted a location at the NW corner of Ajax and Saratoga. But, they were told that space was going to be kept open and they were to be relocated to the SE corner of the same intersection. Additionally, if they were placed there, the station could make McDonalds absorb the cost of removing some asbestos laden piping at their expense thus saving the station money! The changed location turned out to be a poor choice, as cars backed up onto Enterprise around lunchtime trying to go through the drive through. This problem corrected itself approximately six months later as the newness wore off and business leveled off.

The Civilian Personnel Department became the Consolidated Civilian Personnel Office in May, assuming control of the civilian personnel actions for all of NAS Jacksonville and Ceci Field.

On July 3, Florida Governor Bo Graham and Congressman Charles Bennett were present at the NARF for the dedication of two facilities, one designed for cleaning and plating and the other as an aircraft acoustical enclosure (known as the Hush House).

In September, an event that had not occurred at NAS Jacksonville since 1946 took place as a PBY-5A Catalina landed at the station. The dream of Captain Kehrli, the station's Commanding Officer, was to restore the plane and put it on display as a permanent reminder of the major role played by the air station during World War II. The renovation process continued through 1985 until it was completed in the summer of 1986. Captain Kehrli personally participated in all phases of the renovation, as did Wayne White. Mr. White remains at Air Operations today, and has since helped in the restoration of several more of the stations static aircraft.

The "Emerald Knights" of HS-75, a reserve helicopter squadron, made NAS Jacksonville their new home on October 1985. In November 1985, NARF started reworking the F/A-18 Hornet. The F/A-18 program would eventually phase out the Corsair IIs.

1986

The main and Birmingham gates underwent their first major changes since the 1950s in May, in order to better control the reversible lane change times. Also in May, Scout World-Armed Forces day weekend was clouded by warnings from a Public Works employee of asbestos contamination at the planned activity site. The employee had been complaining to the press for months, although base air sampling results never showed any asbestos hazards.

The now fully restored PBY "Catalina" made its final flight with Captain Kehrli at the controls on May 16, 1986. Captain Kehrli had pushed hard to have the aircraft restoration completed before his June 24 change of command. In order to make the farewell flight, he had to receive special permission from the Chief of Naval Operations. On Sunday, June 8, the plane was put on permanent display just nside the Main Gate at the SW corner of Yorktown and Allegheny. The aircraft is on loan from the Naval Museum of Naval Aviation at NAS Pensacola. It was obtained from the Brazilian Air Force and flown to the base.

Captain William J. Green relieved Captain Kehrli in ceremonies at Hangar 1000 on June 24. Captain Green had arrived with a strict "no-nonsense" attitude, but by the end of his tour, he had become one of the more popular commanding officers. One of his first assigned duties (after he played tour guide to actress Brooke Shields) was to phase out the Marines as the station's security force. The Marines had proudly served since their arrival on May 1, 1940. On October 1, the NAS Jacksonville Marine Detachment was formally deactivated. In a symbolic gesture, Captain Green inspected the new civilian guard force and, after satisfying himself of their worthiness for the job, accepted the Marines' statement of relief and ordered the new guard posted. At the same time the switch to a civilian security force occurred, the Security department moved to Building 875, from its home of 45 years at the main gate. The reason Marines were removed from guard duties was a Secretary of the Navy edict that Marines should be more involved with their primary mission as combat trained Marines.

1987

Captain Green was determined to prevent some of the traffic problems and accidents occurring on the base. In February 1987, all new traffic signs were installed as the first of many improvements that continued throughout the 1990's. As a result, the number of auto accidents on base had begun to decline.

The Materials Engineering Laboratory (MEL), a 16,500 square foot, $2 million addition to NARF, was dedicated on April 13, 1987. The NARF had recently undergone its fourth name change and was now called the Naval Aviation Depot (NADEP). Congressman Bennett again served as the keynote speaker, as he had been for so many of the station's building dedications and commissioning ceremonies in the past. The MEL would allow NADEP to explore less expensive materials for performing various rework processes and to reduce the amount of hazardous materials then in use. By this time, NADEP was reworking an

average of 42 A-7 Corsair light attack aircraft and 52 P-3 Orion aircraft per year in addition to 429 engines and 56,000 component units. Work had also begun on the Navy's new F/A-18 "Hornet" aircraft and engines.

On May 2, the Blue Angels performed their first air show in the new F/A-18 "Hornet," at Cecil Field. NAS Jacksonville had wanted to host this air show, but lost out to Cecil Field.

On May 7, it was announced that Apex International Management Services had submitted the winning bid to perform base maintenance. The Government maintenance employees submitted a bid to keep their own jobs, but placed a distant third in the bidding process. Their bid of $25 million was $6 million over the winning bid of Apex. The loss of the contract meant that at least 220 government maintenance workers would lose their jobs. They fought hard to prevent this and their Union attacked by insisting that the contract was flawed, causing the bids to be unfair. Everyone, including Captain Green, was wondering about the ramifications of losing the knowledge and experience of these longtime maintenance employees. It wasn't until April 1988, when the Chief of Naval Operations decided to award the contract to Apex, that the bitter fight finally ended. Captain Green cut short a trip to Norfolk, Virginia, to return to the station to personally sign each termination notice to the affected employees, an unpleasant job which he was not happy about doing. Apex began performing maintenance activities shortly thereafter. Barely ten months later, Apex packed up one Saturday morning and just walked off the base, declaring bankruptcy and accusing the station of breach of contract. The Naval Aviation Depot was particularly concerned, as they had paid $1.2 million for maintenance service, and claimed to have received only $200,000 of actual work in return. The former Apex employees, who had not even received their last paychecks, were immediately placed back on the government's payroll until a new contract for maintenance services could be awarded.

In May the base chaplains were busy providing much needed support and guidance to the families of the victims of the Mayport-based USS Stark after the ship was hit by an Iraqi missile killing several crewmembers. The chaplains often provide such guidance to Navy personnel on a daily basis, but this was an especially trying time.

NAS Jacksonville received the Secretary of the Navy Safety Award in June due to the 44 percent mishap reduction rate. The station had received this award many times in the 1950s, and another trend for safety excellence was starting in the late 1980s. The station would win the award again in 1988.

The new Secretary of the Navy, James II Webb, Jr. held a press conference at NAS Jacksonville in June. This was the first time the Secretary of the Navy had been to the station in many years.

The Jacksonville Navy Flying Club moved to its permanent new home in July, just as a rabies quarantine was instituted on base. The concern for rabies became so acute, wild cats which had continually roamed the station had to be trapped.

VP-24 became the first P-3 squadron to accept an "Orion" with the new tactical paint scheme in October. By 1991, all of the Navy's operational aircraft would have this paint scheme. This scheme was designed in part to make it difficult for foreign forces to determine where a particular squadron was, as no squadron identifying marks are used in this paint scheme, except for the small Bureau number. The paint is also radar absorbing.

The Defense Mapping Agency Detachment, present at NAS Jacksonville since 1941 when it was established as the Air Navigation Office, moved to MacDill Air Force Base on November 30. Vice President Bush landed and held a mini news conference on November 12, as he made a brief stopover on his way to Orlando. Finally, the Naval Air Reserve Squadron formed Attack Squadron 2074 (VFA-2074), the first on the East Coast to use F/A-18's, on December 1.

1988

As 1988 started, over 117 construction and remodeling jobs valued at $105 million were under way. Almost every section of the station was affected by some type of project, especially the now heavy air traffic, as new surfaces were being added to the runways.

Captain Norman W. Ray relieved Captain Green on July 20, 1988. In a break from previous change of command ceremonies, this ceremony was held outside the Officers' Club on the grassy area overlooking the St. Johns River. In his opening comments,

Captain Ray stated "On my drive down, I was thinking of all of the changes that I was going to make to improve the station. So you could imagine my surprise when I got here to find nothing really needed to be changed!" Captain Green, the departing Commanding Officer, wore a Hawaiian lei around his neck in preparation for his next assignment in Hawaii.

Captain Ray, affectionately known as "Stormin' Norman," had a dry sense of humor, and was an excellent administrator for the station. One of the first problems to confront the new Commanding Officer, which could have easily been a public relations nightmare, was a bird hazard on the runway. Recent mowing along the runway had scattered millions of grass seeds, attracting a large number of doves. The birds damaged three aircraft engines and caused two near accidents by the P-3 Orion aircraft. One takeoff even had to be aborted at the last minute due to the winged intruders. Not wishing to harm the birds, Captain Ray decided to try to scare them away. Pyrotechnics were used to frighten away most of the birds, but the "hard core" remnants that refused to leave were still a problem. Eventually mowers with grass catchers were called in and the problem was alleviated after the grass seeds were removed. To the dismay of his many admirers at NAS Jacksonville, Captain Ray's tour was cut short a year when he was selected in June 1989 as the Executive Assistant to the recently appointed Secretary of the Navy. He announced at a morning department head meeting that the new Secretary of the Navy had called him the night before and offered him a job. He called back the next morning to make sure of the previous call, since he had been enjoying an evening "toddy" and wanted to make sure he heard things right! He had, and he announced his decision at the department head meeting.

On 14 December, the new addition to the Naval Hospital was dedicated. The Outpatient Clinic Annex addition encompassed 88,350 square feet and cost $8.9 million. Shortly after it opened, a 1-ton piece of concrete slab hanging from the second floor broke off and crashed through some windows on the first floor, just missing a patient in an office. The Resident Officer in Charge of Construction was called to investigate if a design flaw or faulty construction was the cause. It was finally determined warm weather had caused cracks in the cement that caused the slab to fall.

1989

New construction was the order of the day as 1989 started. A new $13.9 million Commissary Store and Navy Exchange was being built near the east end of the station. Also being constructed was a new $2.9 million Brig that would enable female prisoners to be kept on base for the first time.

On February 22, the main runway at NAS Jacksonville reopened after nine months of repairs. Captain Ray, the Commanding Officer, had seen the runway closed since he became CO. He came to the reopening ceremony wearing a flight suit. He stated: "It's not by accident that I chose to conduct this ceremony wearing the uniform of an aviator… It's an appropriate way to rededicate the heart of this naval air station." The project cost $11.3 million. Traffic had been diverted to NAS Cecil Field during the repair process.

May saw a Navy bias suit finally go to trial. The suit was filed by African-Americans at the Naval Aviation Depot who claimed a system existed where whites were promoted over more qualified blacks. The race-discrimination issue actually started in 1973! Those filing eventually lost the case.

Captain Kevin F. Delaney relieved Captain Ray on June 29, 1989. The guest speaker for the ceremony was Captain Ray's new boss, Secretary of the Navy H. Lawrence Garrett III. Captain Ray had already served Secretary of the Navy Garrett when Mr. Garrett was the Under Secretary of the Navy.

Captain Delaney, a highly decorated Naval officer, came on board with tremendous enthusiasm for his new assignment. The new Captain set some of his priorities as recycling, child care, education, supporting the station's clubs, improving station appearance and anything else that could be done to improve morale for the sailors and their families. One week after taking command, Captain Delaney learned the station was being placed on the superfund list for hazardous waste, bringing the environmental program to the forefront almost immediately.

The new Commanding Officer had his first big challenge on 17 July. The base maintenance contractor, Apex International Management Services, just walked off the job. To keep base maintenance on going, 157 of the former workers were made temporary civil service workers. The company filed for Chapter 11 bankruptcy in March and later filed suit against the U.S. Navy for breach of contract.

NADEP instituted a new program called Productivity Gain Sharing, which went into affect on July 1. The program, which appeared to be working well, gave a percentage of any profits received through increased production back to the employees. The first quarter saw all NADEP employees receiving an extra check for $240.

The huge new Commissary/Navy Exchange complex was dedicated on July 12 and the new Naval Regional Data Automation Center building was dedicated on September 15. Also in September, Captain Delaney established the goal of achieving 100% high school graduates for all personnel stationwide at NAS Jacksonville. His system of first identifying all those who did not have high school diplomas or GED certificates and then referring them to Navy Campus for the GED test had resulted in 18 personnel passing the exam between September and December.

One of the major programs that benefit both the environment and the sailor started in September as Captain Delaney started his recycling program. This program not only reduced the cost of disposing of garbage, but saw increased money going into the Morale, Welfare and Recreation programs on base. It would become the model for all other naval installations.

Some interesting facts from this year are as follows. The base actually made $131,000 profit from the use of pay phones (believe it or not). NAS Jacksonville also had 9,167 military and civilian employees and had a $256.2 million economic impact on Jacksonville. As the decade drew to a close, times ahead were looking a little tougher. Deep cuts in defense were expected in the 1990s. Few, if any, new construction projects were expected to be funded. There was even talk of reducing the size of NADEP at Jacksonville. But initial planning was underway to celebrate the Golden 50th Anniversary of the station on October 15, 1990.

THE 1980s IN PHOTOS

This photo, generated by the Naval Air Rework Facility, shows the first F/A-18 at the facility in 1985. (U.S. Navy)

NCN-121 Super Constellation of VAQ-33 was reworked by the Naval Air Rework Facility in 1981, as a special project. (U.S. Navy)

Safety has always been a priority at NAS Jacksonville, as the awards received in the 1980s show. (U.S. Navy)

NAS Jacksonville main gate showing F11F Tigercat, 1985. The Blue Angel was removed shortly after this photograph. (U.S. Navy)

Station Commanding Officer CAPT Green assists movie star Brooke Shields on a visit to the station, July 1986. (U.S. Navy)

Vice President Bush is welcomed to the station by Commanding Officer CAPT Green and ADM Jesburg, November 12, 1987. (U.S. Navy)

A damaged A-7 Corsair from the May 22, 1981 USS Nimitz accident waits for repair at the Naval Air Rework Facility. Six damaged aircraft were brought to the depot in June 1981 for repair. (U.S. Navy)

The PBY is ready for final paint prior to going on static display on October 24, 1985. (U.S. Navy)

Above: An A-7 Corsair is backed into the engine test cell on the NAS Jacksonville seawall, August 1981. (U.S. Navy)

Right: A reworked engine is carefully slipped back into the fuselage of an A-7 Corsair in August 1981, at the Naval Air Rework Facility. (U.S. Navy)

The wings of an A-7 Corsair are removed at the Naval Air Rework Facility in August 1981. (U.S. Navy)

The first scheduled depot level maintenance is completed on this F/A-18 Hornet at the Naval Air Rework Facility on May 22, 1985. (U.S. Navy)

SH-3 Sea King of HS-5. (U.S. Navy)

SH-3 Sea King of HS-1. (U.S. Navy)

SH-3 Sea King of HS-9. (U.S. Navy)

SH-3 Sea King of HS-3. (U.S. Navy)

This photo, taken on the coldest day ever recorded at the station on January 21, 1985, shows all six assigned patrol squadrons at the station for the first and only time. (U.S. Navy)

SH-3 Sea King of HS-11. (U.S. Navy)

SH-3 Sea King of HS-75. (U.S. Navy)

SH-3 Sea King of HS-15. (U.S. Navy)

VP-62, P-3 Orion. (U.S. Navy)

Below: One of the station-based C-12s flies over downtown Jacksonville. (U.S. Navy)

SH-3 Sea King of HS-17. (U.S. Navy)

Helicopter Antisubmarine Squadron One (HS-1) sailors pose for a group shot in front of an SH-3 Sea King helicopter. Without these people, nothing flies. (U.S. Navy)

Aerial view of Hangar 30, home to the largest squadron in the Navy, VP-30. (U.S. Navy)

NADEP Jax get $70 million boost with Greek A-7 contract, April 9, 1992. (U.S. Navy)

THE 1990S IN PHOTOS

Ron Williamson, station safety manager and CAPT Bob Whitmire hoist the first Secretary of the Navy Safety Flag at the station's front gate, July 1996. (U.S. Navy)

Medal of Honor recipients Thomas Hudner, James Williams and Donald Ballard stand as CAPT Whitmire dedicates Patriots Park. (U.S. Navy)

Fire damage to Building 13 which burned March 12, 1993. (U.S. Navy)

CAPT Woodledge, CO of the Blue Angels for their 50th anniversary, and CAPT R.M. "Butch" Voris, first Blue Angel, at NAS Jacksonville 1996 Air Show. (Author)

Vice President Dan Quayle shaking hands with CAPT Kevin Delaney, July 9, 1990. (U.S. Navy)

T-45s on the line at NAS Jacksonville, August 1998. (Author)

NADEP jet engine facility. (U.S. Navy)

50th anniversary of NADEP's first aircraft rollout, August 14-15, 1991. (U.S. Navy)

A-7 Corsairs lined up for rework on November 21, 1991. These aircraft were reworked and sold to Turkey. The Navy phased out the A-7 in August of that same year. (U.S. Navy)

Poster advertising soccer match between U.S. and Soviet Navy teams played at NAS Jax on July 17, 1991. The U.S. won 5-1. (U.S. Navy)

Group photo of NAS Jacksonville team receiving award at the Pentagon, May 24, 1991. (U.S. Navy)

First F-14 Tomcats arrive at NADEP for rework on January 6, 1994. (U.S. Navy)

Congresswoman Tillie Fowler, member of the U.S. House of Representatives, stands beside "her" helicopter at NAS Jax. Note the torpedoes loaded on her port side. (U.S. Navy)

CAPT Turcotte gives a thumbs up as he flies in the Navy's last A-4 Skyhawk.

NADEP Jax T-2 line in Hangar 140, June 1993. (U.S. Navy)

Navy personnel spell out Jay Beasley, for the hangar dedicated in his honor. (U.S. Navy)

NADEP Jacksonville advertises the fact they were the home of the new Navy stealth "Avenger," October 1990. This aircraft was cancelled by the Navy shortly thereafter. (U.S. Navy)

CAPT Turcotte begins a golf course expamsion project in 1999.

Helicopter Squadron 3 was the first to recieve the SH-60 Seahawk, the replacement for the SH-3 Sea King, in 1991. (U.S. Navy)

Station Search and Rescue (SAR) helo, NAS Jax. (U.S. Navy)

P-3s on the ramp showing a mix of the old (white) and new (grey) color scheme, March 1994. (U.S. Navy)

VP-45 P-3 Orion prepare for a night takeoff at the station. (U.S. Navy)

HS-5 HH-60H helicopter deploys Marines to the carrier deck with a fastrope. (U.S. Navy)

A VP-30. P-3 Orion lands at the east runway approach. (U.S. Navy)

One aircraft of all six station-based Viking squadrons fly over the station, January 1998. (U.S. Navy)

VR-58 Command photo. The aircraft, a C-9B Skytrain, will be replaced, starting in early 2002. (U.S. Navy)

VP-49. (U.S. Navy)

HS-75 receives the first SH-60 Seahawk, December 1999. (U.S. Navy)

The last SH-3 Sea King, belonging to HS-75 is prepared for a final flight on December 1, 1999. (U.S. Navy)

Aircraft types assigned to NAS Jacksonville in 1997. (U.S. Navy)

1990

Just as the 1940's were defined as a period of unprecedented growth, the 1990's came to be symbolized by terms that all meant just one thing - cuts. The first one that was heard throughout the 1990's was BRAC - Base Closure and Realignment Commission. This commission would eventually recommend the closing of NAS Cecil Field, which also had a major impact on NAS Jacksonville. The Naval Aviation Depot spent considerable time during the 1990's defending itself against closure, as did NAS Jacksonville proper. In addition to this commission, other actions were being taken to reduce squadrons, personnel and budgets. The 1990's could be considered the decade of squadron disestablishments. (Ships and Navy activities are commissioned and decommissioned, but squadrons are established and disestablished.) The station saw seven squadrons disestablished during the decade. Other buzzwords were intermingled throughout the decade beginning with TQL - Total Quality Leadership; consolidation - as in consolidating functions within base fencelines to reduce costs; "thinking out of the box;" "doing more with less;" and the one that has probably caused more studies, paperwork and confusion that anything - regionalization.

Captain Delaney was at the helm of the station as 1990 started. He had ambitious plans for the stations' Golden 50th Anniversary Year. The station was going to celebrate with an air show, the first to be held at the station in some 17 years. There was a steep learning curve for station personnel in putting on the show, and the Blue Angels, who were formed here 44 years earlier, almost were not invited in time for the show. If it were not for the efforts of Maynard Cox and myself to inform the XO and Captain Delaney the air show request for the Blue Angels had to be at the Chief of Naval Operations in two days, and still had to go through the FAA (Federal Aviation Administration) for approval, the invitation would not have made it in time. It took "walking through" the request immediately by Air Operations personnel to make the deadline. Enough

cannot be said for the efforts of Maynard Cox on behalf of the air show. An initial meeting at Congressman Bennett's Office was held to get his support of the planned activities, which Captain Delaney attended.

One of the first messages out of the Secretary of the Navy's Office in January 1990 was to impose an immediate freeze on the hiring of civilian employees. On January 31, Captain Delaney changed the name of the street in front of Building One, the Commanding Officer's building. It was originally named Patoka, after the seaplane tender. In keeping with the station mission of Antisubmarine Warfare (ASW) and for the recent new designation as a Master ASW Base, it was changed to ASW Lane. (The official maps were never changed at the Southern Division Engineering Command in Charleston, however, and Patoka is actually still officially listed today.)

Vandalism was on the rise again in February at the enlisted barracks and at the Golf Course, of all places. Some $50,000 in damage was done to the golf course equipment. A curfew was imposed on children on station for the hours of 11:00 p.m. until 6:00 a.m. to try and curb the damages. A new program for civilian employees was also announced on February 27 for On-The-Spot-Awards. If an employee did something significant, an immediate award of between $25 and $100 could be given. It remains at the station today, but was stopped temporarily in January 2000 due to lack of funds. The NADEP shops were busy turning out fleet ready FA/18 Hornets, but converting one for use by NASA was a first.

Thursday March 15 saw the first meeting to plan the events of the 50th Anniversary air show. I will always remember this meeting as one of the most unique in the station's history. Two commanders got into an argument right at the start that had to be settled in the Executive Officer's office, as the rest of the committee waited for over 20 minutes! Also in March, Apex International, who had walked out of the base maintenance contract the year earlier, was filing a $6,000,000 damage suit against the Navy. A new contract had been awarded the previous October to Flour Daniels, who became an

excellent base maintenance contractor.

Captain Delaney was also emphasizing his goals for the year in a message in March. Recycling had been going on for only six months, but the station had already recycled over 300 tons of materials. Benefits to the station were to make the first year's goal of $250,000 in funding. Security was increased and a "no guns permitted on base" policy was emphasized. Base clean up was progressing to the point where Captain Delaney was claming NAS Jacksonville "The Cleanest Base in the Navy." Education of the sailors was also emphasized. Over 3,000 students from the station were enrolled in on-base college classes. His plans for the future were just as ambitious. A new recreational vehicle park was being constructed; a Veterans Memorial was to be placed in front of Building One and a new flag pole added at the main gate. Improvements were being made at the Child Development Center, golf course and at the on-base clubs. We were also trying hard to find an old F6F "Hellcat," the first plane used by the Blue Angels, to restore and place at our main gate. (We eventually gave up when Navy Museum personnel told us of the 14,000 Hellcats built for WWII, only about 12 exist in the world today!)

Senator Bob Graham was having a work day at the NADEP in early April, just as a new warehouse was completed along Roosevelt Boulevard to be used by the Naval Supply Center. As the station was looking at costs, the fact that $12.8 million was spent on utilities gave Captain Delaney the impetus to start an effective energy control program for the station.

May 7 saw the dedication of Manatee Park on the station. It was the first park paid for by using the recycling funds that were being generated. The annual Scout World came May 19 and 20, along with the hot air balloon show. This show was a highlight of the Scout World event.

June 4 saw Pelican's Perch open, making NAS Jacksonville the largest Child Development Operation in the Navy when combined with the other existing facilities. Pelican's Perch was originally designed to be a furniture store for use by the Navy Exchange, but the mission was changed to meet

Captain Delaney's new visions. Child care was removed from this facility in the late 1990's and the building today exists as the station post office. The beginnings of the static aircraft display area were also coming to fruition in May. Only a single PBY aircraft existed as a static display. On Friday, June 15, a P-3 Orion and an SH-3 Sea King helicopter were dedicated.

The new additions to the Paul Nelson Helicopter Training facility were dedicated on July 6. What was only a small helicopter training facility was increased by 80,200 square feet adding 17 classrooms and $136 million in Aircrew and Maintenance trainers. Cost of the new construction was $8.8 million. The main speaker, as at so many of the station dedications, was Congressman Charles Bennett. A plaque honoring Commander Paul Nelson was also unveiled. Three days later Vice President Dan Quayle visited the station. He only remained for a short visit as he attended a fund raising event in Clay County. The new flag memorial at the main gate was dedicated on July 19. At the ceremony along with Captain Delaney were the members of American Legion post 137, local station sailors and civilians and a group of WAVES from WWII who just happened to be at the station for a reunion. The new 40' high flagpole was the first of three planned memorials dedicated to veterans. Post 137 members led by Buddy Ellis, who worked at NAS Jacksonville Air Operations, raised the funds for the memorials.

Visions of future operations in the Gulf in answer to Iraqi actions were having an effect in August. Some 350 workers at the NADEP were being called in on August 11 and asked to speed up repairs on aircraft. NADEP Commanding Officer, Captain David Wynne stated "We've stopped working on lesser priority things and started working on things that were more important," per a Times Union Article of August 12. The priority items were three additional FA-18's; four A-7 Corsair's and jet engines. An unfortunate sidepiece to the NADEP artisans working harder, with considerable overtime, was they almost did too good of a job. They actually turned out so many aircraft and engines that there was a chance they could work themselves out of a job as incoming work did not match output! Doctors, nurses and corpsman were also leaving the Naval Hospital for the Middle East as tensions increased. Iraq had just

invaded Kuwait on August 2, and the station forces were wondering how long it would be before they were involved. Another concern is what effect would this conflict have on the October 15 air show planned for the station. With the planned heavy involvement of military aircraft, would it have to be cancelled?

August 12 saw a program initiated by Captain Delaney that had been asked for by civilian personnel for years. I left the U.S Army Corps of Engineers in 1982 to come to NAS Jacksonville. The Corps of Engineers had initiated flexible working hours, and I thought it would be good for the employees at NAS Jacksonville who worked the standard 40-hour workweek. It was turned down everytime I suggested it, but Captain Delaney finally initiated an alternate work schedule. It allowed the civilian workers to work nine hour days Monday through Thursday, and one eight hour Friday. The other Friday you got off. This highly successful program was initiated to save energy, but its major contribution was to employee morale. It continues at the station today.

September 24 saw the last scheduled depot level maintenance on an A-7 Corsair for the NADEP. This had been the longest running rework for one aircraft type in the history of the facility, and it was coming to a close. The Navy was phasing out the A-7 Corsair's in favor of the F/A-18 Hornet. The "A" letter in F/A actually meant attack, the role the A-7 Corsair was designed for. All employees of the NADEP were invited to the final ceremony. A week earlier, the Armed Forces Softball Championships were held at the station's softball complex.

The month of October would be one of the biggest in the station's history as far as celebrations. With the start of fiscal year 1991 on October 1, another command was immediately established. The start of the fiscal year is traditionally a time when a new command is formed, as a quick look at the station's history will show. There had been a communications department associated with NAS Jacksonville as a department almost since inception. This department, under the Commanding Officer, broke away in 1983 and became a separate tenant organization called Naval Computer and Telecommunications Station. No longer would they be attending the biweekly department head meetings! They then merged with the tenant Navy Regional Regional Data Automation Center (the computer folks) to form one new activity. Also opened

on October 1 was the new ITT/SATO (scheduled airline ticket office) building. This office allows military and civilian employees a place to schedule leisure travel and buy tickets at a discounted price to Florida's attractions and theme parks. The Navy Ball was held on October 7 with Secretary of the Navy Garrett in attendance. He also visited the station the next day.

The other two parts of the three-piece Veteran's Memorial were dedicated on October 13 as the 50th Anniversary Celebrations commenced. Two plaques were placed in front of Building One where they remain today. But the air show was the culmination of the years planning efforts. A highlight of the show, in addition to the Blue Angels, would be a Stealth Fighter, the first to be seen by the public in Jacksonville. This fighter almost did not make the show. Three days before the show, the Air Force called to say the fighter could not make it, as they needed to hold resources in case needed for the Gulf conflict. Captain Delaney made a call to Congressman Bennett, and suddenly the plane was made available in a very accommodating way! As it landed the first day of the show, it brought the (crowd estimated at over 100,000) to a silence as all watched the black chute come out the back of the fighter to slow it's landing.

This show had all the hype of the 50th Anniversary, but the fact NAS Jacksonville had not put on an air show for over 17 years made for a learning curve that was showing. A decision to ask for a "contribution" to park caused some negative comments to be made, and the Navy came out with a policy to stop this for any future air show at a Navy facility. The controllers were extremely busy in the tower just keeping the event flowing as planes for certain acts would finish early and other aircraft couldn't get their engines started on time. Although the show seemed to flow well to the public, hard work was done to coordinate events at the tower!

The static display area was also less than stellar, with a sparse display as compared to the later air shows the station would have in the 1990's. But the public was awed and impressed as the show went on. The first day of the show was dedicated to Congressman Charles Bennett. The second day of the show was dedicated to Alexander Brest. On the second day (Sunday) the Executive Officer (XO) was on his way to have me paged (being the unofficial station historian) by the air show announcer when he ran into me and said they needed a bio on Mr. Brest. "I think he gave some land for the station to be built on, or something, so see what you can find out. We need something for the announcer to read!" were the words of the XO. Being not quite sure where to dig up a bio on someone I only vaguely had heard of, and that was only from being at the Alexander Brest planetarium in Jacksonville, I grabbed Maynard Cox. As we thought, someone had the idea of calling the Florida Times Union. Maybe they had an archive of biographies. We called (on a Sunday morning), and low and behold they were most helpful when we explained our dilemma. He read what they had and I took notes as fast as I could, only hearing bits and pieces as jets flew overhead. Mr. Brest arrived as I was writing a short bio for the announcer to read. I handed it to the announcer, and he introduced Mr. Brest, who was now in the VIP area meeting Captain Delaney. Just as he started reading the bio, an F-86 Sabre jet started a fly-by, and not a word could be heard by the audience. He finished just as the jet noise was subsiding! I was actually relieved, as we had totally missed the mark as to why Mr. Brest was being honored. I always assumed someone must have known something, or he would have not been invited, but whom? The Times Union person giving us the bio never mentioned how Mr. Brest was responsible for so much of the construction of the base in the early 1940's. So I had just thrown into the bio he had donated some land to make some tie-in, which of course he had never done!

Of particular note, I found out years later, were the hurt feelings of the original pioneers of the station, the reunion members of squadrons VN-11, VN-12, VN-13 and VN-14. Members of these early station squadrons had also planned their reunion in Jacksonville for the same weekend. They asked the air show coordinator for a space so they could sit and watch the show, and he *denied their request*. They were welcome to come and watch the show along with the rest of the general public, but no special seating was to be given for this small group. Here were the aviators who formed the first squadrons, fought in WWII and were instructor pilots here during the war, and their request for a special seating area *was denied*! Captain Delaney had not gotten word of their request or the denial or he would have immediately corrected this situation.

Having just completed the first book on the history of the station, I did not even know this reunion group existed, and actually missed most of what they accomplished here in that first publication. I went to their reunion in Pensacola the next year and have attended a couple of their reunions since then. But it took a while for them to get over their hard feelings and start telling me the history of this station in WWII, as I tried to apologize on behalf of the station. Much of the information on aviation activities and early station history we now know today has come from their donated photographs, letters, stories and memorabilia.

The highlight of the events for me was the luncheon on October 15. Class 12A-40-J had been formally invited to the station and was to be honored at a luncheon. The members of this group were the first class to receive their wings at NAS Jacksonville in 1941. Marine Corps Lieutenant General Robert Keller, who had trained at the station in WWII and went on to become a decorated Marine pilot, was the principal speaker. Most of that first graduating class had made it through the war and attended this event. This busy month was finished with the station receiving it's third environmental award since Captain Delaney's programs had started.

1990 finished with the nation and station personnel still watching the events in the Persian Gulf. It had been a remarkable year in the station's history that would be hard to match!

1991

Budget issues again started out the year 1991. Captain Delaney began the year with a memorandum announcing that significant budget restrictions had forced janitorial services to be reduced to two times a week in offices. The NADEP, who did not initiate an alternate work schedule when NAS Jacksonville had the previous August, decided to try it with the new year. The traffic signals were reprogrammed to handle the major traffic changes. Although the program there was as popular as it was with the NAS Jacksonville employees, NADEP cancelled the program after finding few savings and an actual slow-down of production occurring with the one lost work day every two weeks.

The first accident occurred at VP-24 in their hangar the first week in January.

The hangar doors were shut and they snapped off the MAD boom (the long tail) off of a P-3 that had not been brought completely into the hangar! The big news, however was the U.S. attack on Iraq. The events started on 17 January and personnel all over the base watched the events unfold live on television. Of particular concern was the report of a lost pilot from NAS Cecil Field. Lieutenant Commander Michael Spiecher became the first Gulf Coast war pilot missing in action, later presumed dead. A total of five personnel from NAS Cecil Field would be killed in the Gulf War.

The morning of January 18 saw an unprecedented traffic jam of cars trying to get into the station off of Roosevelt Boulevard as security was tightened and every identification badge was checked at the gate, a slow methodical process. My normal 15-minute trip to the base took over two hours. The executive officer, who heard complaints from almost every city and county leader in Duval and Clay counties, told the security officer to not do this intense check the next day. Station personnel were satisfied to be late, as long as they were given administrative time for being late. The next morning saw the exact same traffic backup again, as the word was not passed down through the station security chain! Even a frustrated security officer waited in traffic trying to get to the base to tell his gate guards to knock it off. After a second meeting with the Executive Officer, the word was absolutely passed down for the third morning and traffic immediately got to a more normal state! Patrols of the stations vulernable shore along the St. Johns River were also started. It seemed much of the publicity concerning Desert Storm passed the station by. NAS Cecil Field's fighters were deeply involved, as were Naval Station Mayport ships. NAS Jacksonville's participation consisted of the helicopter squadrons (HS-3 and HS-9) flying rescue missions and bagging some mines and POW's. Squadron VR-58 transported men and equipment. Even squadron VR-58 did not serve quite as important a role as they could have, being kept in Italy for most of the conflict.

February started out with the announcement that VP-56 would be disestablished at the station, as the cuts continued. The squadron had been at NAS Jacksonville since 1972, and the loss meant eight P-3 Orions, 60 officers and 250 personnel would be gone.

March 1 started out with the renaming of Hangar 1000. The hangar was the home to all of the P-3 patrol squadrons since it was completed in 1971. The hangar was renamed for Jay Beasley, Honorary naval Aviator number 11. Although a civilian test pilot for Lockheed since 1943, he had flown with over 5,000 Naval Aviators. He had over 9,400 hours in P-3's and was actually known as "Mr. P-3." On the 6th of March, Captain Albertolli took over as Commanding Officer of the Naval Aviation Depot. The depot could not have used him more. His business savvy and leadership skills would single-handedly save the Naval Aviation Depot from closure. This will be explained further as his initiatives are introduced in this writing. The phrase he most often used during his tour was: "Do the right thing!"

In a message to the station dated March 29, it was announced NAS Jacksonville was the selection for the 1990 Commander in Chief's Installation Excellence Award. This award is the highest any station can receive. Every year the stations and bases from all the services have nominees and one station is selected from each service as the best. NAS Jacksonville was the best base in the Navy. Coming with the award would be a check for $100,000 the Commanding Officer could spend as he saw fit. An interesting article being circulated in March stated P-3's were the most lightning struck aircraft, by far, that the Navy had. Although few problems had been reported with the station's P-3's, being in the nation's lightning capital, the report was read with interest. March 27 saw Whitney Houston visit the station. After receiving an honorary flight suit, she was escorted around the station by Captain Delaney.

May 1 saw an experiment with the announcement an express galley would be started. This would allow military personnel to call in orders and have food ready for quick pick up. A delivery truck was added after this programs initial success. A traffic light was removed from the corner of Langley and Saratoga on May 17. A traffic study by NAS Safety determined a four-way stop would better serve traffic.

May again saw the 11th Annual Scout World/Armed Forces Day event, with the balloon lift-off the big attraction again. May 23-24 saw a planeload of station official's head to the Pentagon to receive the Installation Excellence Award. As they were receiving the award in Washington, a ceremony at

NAS Cecil Field was phasing out the A-7 Corsair. Squadrons VA-46 and VA-72 were the last ones to fly the attack plane. Both A-7's flown in this final flight had served in the gulf war.

June 1 saw an announcement the Naval Aviation Depot was to be added to a list of facilities to be considered for closure. Captain Albertolli had already been moving in early May to save the facility, however. Initial planning was being made to cut 700 jobs at the depot. The depots at Pensacola and Norfolk were also announced as being potential victims for closure. But personnel there were sure *they* would not go! Pensacola worked helicopters and Norfolk was too close to the Navy's main political structure. But NADEP Jacksonville had one thing they did not have - Captain Albertolli. He knew costs had to be cut and the depot had to show a profit, or operate in the black. All that was needed for an official announcement was approval from Secretary of the Navy Garrett.

June 28 saw the official disestablishment ceremony for VP-56. The "Dragons" relocated to NAS Jacksonville in 1970 and after an hour-long ceremony, were no more. The squadron commissioning pennant, a long naval tradition, was kept by the last commanding officer, Captain Rush E. Baker, III. Following right on the heels of that disestablishment was the ceremony on July 2 of HS-17. Of all the ceremonies I have attended at NAS Jacksonville, this was one of the finest I have ever witnessed to this day. Instead of disestablishing a squadron that was formed at NAS Jacksonville in 1984, it seemed like a death in the family! Mentioned in the final ceremony was the fact the squadron should have received the station deploying squadron safety award. I had just started that award the previous year as a way to get the squadrons more involved in the station's safety program. They were tied with another Helicopter squadron in our awarding criteria for first, and I do not remember today what made the final decision. But after hearing at the ceremony how important this award would have been, in retrospect I should have given it to them!

On July 1, the official announcement was made to the press that, in fact, 700 of the 3,300 jobs would be cut at the depot. The loss to Jacksonville would mean $60 million if all 700 lost their jobs, not to mention what the employees felt about loosing their jobs! The Base Closure and

Realignment Commission also voted on the second round of base closing for 1991. The Navy facilities in Jacksonville had once again been spared. July 17-18 saw an event thought impossible just a few short years earlier. Three Russian ships docked at Naval Station Mayport for a visit to Jacksonville. The ships were the guided missile cruiser Marshal Ustinov, the guided missile destroyer Simferpol and the replenishment ship Dnester. Tours of all station planes were set up for the visitors, but the big event was a soccer game between the two super powers at the station. In a rough played game, the Americans won! The real highlight of the Russian visit was the interaction allowed. The Russians were eager to trade anything they had for American souvenirs, and base personnel were trading handily! The leader of the Russian delegation presented a model in a diorama case of an American Brewster Buffalo with Russian markings from WWII. America was phasing out the older planes as being obsolete. NAS Jacksonville could not scrap them fast enough during the war. But Russia desperately needed aircraft, and they were given to them. The presentation of the model was accepted by the Executive Officer. On the politically correct front, an official memorandum from the executive officer on July 19 directed all station personnel to cease answering the phone "...May I help you, Sir?" or "May I help you Sir or Ma'am?" to a gender neutral greeting of "May I help you?" As our new official phone greeting was being signed, Naval Hospital Jacksonville was having a 50th Anniversary celebration of their own. A ceremony and picnic attracted 1,000 people to Orion Park on station. NAS Jacksonville was also receiving yet another award for recycling. The station had collected 500 tons of office paper, 475 tons of cardboard and 240 tons of newspaper. It had also saved the station $200,000 on reduced waste hauling.

August 1 saw 700 workers at the NADEP receiving personal letters telling them whether or not they would lose their jobs. Some opened them immediately; some took them home to open them with their families. Another 270 were to retain their jobs, but be demoted. A planning notice was also signed for the change of command ceremony for Captain Delaney. The station would lose him on August 26. August 15 saw a 50th Anniversary ceremony at the NADEP. I had gotten my wife, who worked at the NADEP, to ask if they wanted a Stearman, the first aircraft to ever be reworked at the NADEP, flown in to celebrate the 50th Anniversary of that event. Captain Albertolli loved the idea, and thought it might bring a little relief from the recent personnel cuts. A Stearman was flown in to the station with Captain Albertolli on board for the flight. He had the plane parked in their main hangar and allowed personnel to get pictures taken with the plane. A ceremony was held with coffee and cake for all that attended. The highlight for me was to be able to fly back to Herlong from NAS Jacksonville on the Stearman!

Ground was broken for the new $2.1 million building to house enlisted corpsmen from the Naval Hospital on August 20. It would be named Hancock Hall, in memory of 2nd Class Scott Hancock who was killed in Viet Nam trying to save wounded Marines. The galley at NAS Jacksonville had actually been named Hancock Hall some years earlier, but it was felt this facility actually served his memory better. The Hancock family was in attendance. On the 21st, plans were being made for a park on the property owned by the Navy west of the station by the Trout River. Developers would routinely call the base wanting the property to develop, but a park was in the planning now. The news was also a little better for the NADEP employees who had received reduction in force notices. Some 330 would be hired back temporarily.

On August 26, Captain Cramer relieved Captain Delaney. It had been a remarkable tour. Captain Delaney had been selected to the rank of Rear Admiral, an incredible feat showing just how strong his accomplishments were. Promotions to Admiral are rare, and even rarer for someone who was not on the front lines during Desert Storm! He had, however, flown almost 700 combat missions in Vietnam! Listed in the official change of command program were some 15 major awards the station received during his tour. This feat has not been even approached since. For his remarkable progress in recycling, a line of recycling trucks drove by as a salute, although the Times Union headline stated "Navy chief "saluted" with trash." Captain Cramer came with a wonderful sense of humor and at the change of command stated "I know I have big shoes to fill, but I also found Captain Delaney and I have the same size shoe!"

In September the first SH-60 Seahawk helicopter arrived on station. This helicopter would eventually replace all of the SH-3 Sea Kings being flown by the stations helicopter squadrons.

On October 10, an SH-3 Sea King attached to HS-11 crashed into the Atlantic during a fleet exercise off the coast of Bermuda. All four members of the crew were killed. Wreckage was found almost immediately after the search was started. A sudden catastrophic impact with the ocean was thought to have occurred.

November saw major road refinishing commence. Yorktown Avenue was completely repaved, a job that was sorely needed. Other roads had markings added. On Friday, November 15, the Air terminal in building 118 was dedicated to the memory of Winton "Buddy" Ellis. Mr. Ellis was a long time, well-liked employee of the Air Operations department. Captain Delaney was the guest speaker. The day before, Vital Signs had their grand opening celebration. The club was touted as being a high-energy night club. It was initially a very popular club, but trouble with certain theme nights started a slow demise of activities, and profits. Captain Cramer was also facing yet anther round of budget cuts. He had to cut $4.2 million of his $27 million budget. Proposals included not opening the galley, which had just finished a $1.2 million renovation, reductions in civilian personnel, removal of telephones and other initiatives. Fortunately, in a message from Captain Cramer on December 18, a furlough of employees for 22 days to save funding did not have to be instituted. Captain Cramer finished out the year with a message to all personnel on the base giving his overview of operations to date and his goals for 1992. The station was still winning awards and still moving in the right direction!

1992

January started out with the NADEP announcing another 200 positions would slowly be phased out during the rest of fiscal year 1993, as the cuts in defense budgets continued. Additional cost cutting measures had alleviated the necessity of having to furlough some 400 civilians at NAS Cecil Field and another 450 at NAS Jacksonville. Captain Cramer stated there was still work to do to save funds, however. He expanded the alternate work schedule to include military personnel so buildings could be closed for three-day weekends to save on energy costs. Coming to the station in early January was Pat Dooling, the new Public Affairs Officer for the Admiral's staff and dual-hatted to the Commanding Officer, NAS Jacksonville.

Having been the safety officer for the station since 1982, I can recall many memorable mishaps, mainly due to the way they occurred. The one that occurred on February 12, however, will be one I will remember for a long time to come. The helicopter squadrons have personnel in front of every helicopter flight who direct the helicopter as it taxies along the flightline area. The helicopter ramp/flightline are also along the St. John's River. The safety officer for HS-7 was the pilot of a flight taxing along the flightline, and a member of the squadron who was waving his wand and walking backwards, was directing him. Sure enough, he walked backward right off the end of the ramp area and fell into the St. John's River. The pilot and co-pilot were of no use to him, as they laughed uncontrollably. He swam back to a seaplane ramp leading from the river and came walking back up, unhurt, except for his pride. After the mishap, a wide yellow warning line was painted the length of the seawall area to provide a warning for future backward walkers!

I had received a call one afternoon from someone asking if the station would like a P2V Neptune patrol plane for the static display area. Mr. Cox and myself had already been working on plans to expand the park, and the addition of a plane that flew at NAS Jacksonville for some 17 years was very appealing. After working with the National Museum of Naval Aviation in Pensacola, a deal was reached with the owner. The plane was to be flown to the station from Kissimmee, Florida. The press was invited to attend the 3:30 p.m. fly-in on 5 March. I went to the control tower area to watch and Captain Cramer met the Admiral Walker on the flightline, along with Mr. Cox. The plane arrived and flew along the runway for one brief pass, accompanied by the station C-12 utility plane that had flown along as an escort. On the second pass, the landing gear was lowered, and low and behold, the left landing gear did not come down. It was not time to panic, as a manual control installed inside the plane could be used. Commander A.J. Jackson, who was the Air Operation Officer, was on board as an aircrew (observer), and not with an FAA ferry pass! (The pass was for a one-time flight to the station for the Museum pilot and co-pilot only.) He had flown in P2V's, and wanted to make that last flight. Station officials knew, but he was not on board with "official" authorization. Being a big man, "AJ" knew how to manually lower a landing gear and proceeded to do just that. He sat down in an area behind the cockpit, put both feet on a bulkhead and started to turn a crank to lower the main left gear manually. Just as he started to turn the crank, both feet went through the bulkhead! The gear was now not going anywhere. Another pass was made by the station. On the next pass, the pilot would land anyway. This was complicated as the port engine also stoped on the final pass! The landing turned into a one wheel up and one engine out exercise, in front of the press! This was not the way we wanted to remember the old P2V's arrival!! In an incredible job of flying, he touched down on the right main wheels, and flew along until his airspeed dropped and he slowly swung the airplane around on the tip of the left wing. Virtually no damage was done to the aircraft or the port engine, which had been feathered. There were no injuries and the plane ended up in a grassy area and off the runway so as to not interfere with air traffic. There were a lot of nervous folks on the ground, but everyone went for a celebration to the officers club afterwards. An investigation was done that concluded the plane landing gear system should be fully checked out prior to anymore flights. (Like that was a revelation to anyone!) The pilot had been notified not to raise the landing gear on take-off, but it was done for fuel efficiency during the flight hoping it would come down appropriately prior to landing. King Hall was demolished by March 27. The Morale, Welfare and Recreation Department mainly used the building for basketball and youth events.

During this time period, damage was still being done due to the Tailhook scandal. The Secretary of the Navy had just resigned his post, and NAS Jacksonville had to inquire from every military member on station if they had participated in the Las Vegas event, yet again. Not having aircraft assigned at the station with "tailhooks," only a couple of personnel were found to have attended the event and they were nowhere near any of the events that caused the Navy trouble. In fact, the

personnel of Helicopter Squadron 1 actually provided safe refuge to some of the females at Tailhook. Because of the Tailhook troubles, planning was being made for a mandatory "Mutual Respect" standdown for August. All civilian and military personnel of NAS Jacksonville would be required to attend one of the three-day events.

May 16 saw the annual Scout World event return. Absent for the first time in years were the hot air balloons. The event was successful, but drawing smaller crowds than in past years.

June saw the decommissioning of the Mobile Maintenance Facility Detachment Alpha group. This detachment consisted of a group of trailers that could be packed up and moved anywhere in the world on short notice to provide electronic maintenance support. They were located inside the flight area just east of the Patrol Hangar. The station's main entrance brick sign was moved on June 16. It had been in place sine the 1950's, and was scheduled to be demolished. But a petition started by the maintenance director, Mr. Roger Willi, convinced Captain Cramer to relocate the sign to a location where the static display aircraft are located today, on Yorktown Avenue. A new entrance sign was also installed.

August 4-7 saw the Chief of Naval Operation mandated Mutual Respect Standdown. This event was mandated as direct fallout of the Navy's Tailhook scandal. In an open letter to NAS Jacksonville personnel after the event, Captain Cramer noted over 8,000 personnel from 25 commands were educated and enlightened during the very successful program on "Harmony and Human Dignity." My office got additional fallout over the event; we lost control of the base auditorium. We had been given control of the auditorium in 1988 when Captain Ray was Commanding Officer, through a strange twist of events. The auditorium was actually the old base theatre. It belonged to the Morale, Welfare and Recreation Department (MWR) at that time. MWR was not showing movies anymore and so did not put any funds into the facility. Captain Ray went to give a presentation one afternoon, and noticed only one light was working inside. When told the story of how MWR only funded things they make money from, he transferred the building to NAS Safety, since it was mainly used for safety standdowns by squadrons anyway. Once I inherited the building, I decided to fix it up some. One

thing I wanted to do was to place old pictures of the station on the lobby walls, after we renovated it. As I dug into the station's history, I discovered there was little written on the station's past, and that it was filled with significant historical events. In addition, I discovered the 50th Anniversary was approaching, got very interested in the history of the station, and decided to do a book on the history of NAS Jacksonville. We found some old pictures; renovated the lobby as a self-help project; painted the inside; had a new roof installed; had the asbestos lagging removed; and got funding to replace the then 45 year-old movie screen and had some sound equipment installed with a permanent microphone for speakers. Up to that time, anyone using the auditorium had to bring his or her own microphones and speaker equipment. Having but a small budget, we bought minimal equipment that would work fine for your normal sized squadron. The theatre now filled to capacity, however, overpowered the minimal equipment. The Admiral, kicking off the agenda for the Mutual Respect Standdown could not be heard, and he ended up shouting his opening remarks. It was shortly thereafter I received a call from the newly established Command Services Department notifying me that the Executive Officer had transferred custody of the auditorium to them. (There was also a better sound system installed almost overnight!) Chief of Naval Operations Admiral Kelso visited the station on August 14 and Secretary of the Navy O'Keefe visited on September 10.

Although we had just finished training due to the Tailhook scandal, the Tailhook investigations were just starting to roll along. A letter from the Naval Legal Service Office announced all aviators in the Jacksonville area who attended the Tailhook convention would be interviewed by an investigation team the last week in September. Defense lawyers would also be detailed to any officer who thought they might want to discuss the facts of their particular case.

October 1 saw three commands change names. The first to change names was the Admiral's staff. Commander, Helicopter Wings Atlantic became Commander, Naval Aviation Activities Jacksonville. It was originally supposed to be Commander, Navy Jacksonville Aviation Activities but was changed when the acronym was discovered to be too close to COMNAVJAXAAS, and that did not sound too good! This was not the first time an activity had to change their name due to the acronym. The NADEP, in the late 1960's, was close to being called the "Fleet Air Repair Terminal" until it was discovered this was shortened to FART! The Consolidated Civilian Personnel Office changed its name to the Human Resources Office and the Public Works Department became the Public Works Center, a separate command. It was mostly a name change for the first two, but adding the Public Works Center made a big change. The Public Works Center was supposed to assume all maintenance, engineering and transportation issues for the three local bases. The Commanding Officers, concerned about that loss of engineering and maintenance control, however, kept existing staffs. This "double staff" would exist until regionalization efforts would bring them together sometime in FY 2000. One of the more infamous speakers who came to the station since I had been there was Oliver North, who spoke to a packed auditorium on October 19. He gave a stirring speech and afterwards was inundated for requests to pose for pictures and sign autographs.

The Blue Angels performed at the station October 30 through November 1. This show was noted for the arrival of the Stealth fighter. It arrived early the Thursday before the Friday practice. Anytime the Stealth is present, it has to be guarded in a secure hangar. Hangar 117 was to be the location for this show. The plane was taxied and brought right into the hangar. As it turned near the hangar entrance, the jets faced the open hangar and the heat from the engines caused the entire water fire extinguisher system to start spraying water. The hangar was flooded by the time they could be secured! The show went on without a hitch, and was run a lot smoother that the 1990 show, now that the station had some experience and corporate knowledge behind it.

December finished out the year with the bike patrol started at Security. This patrol would allow savings on vehicles, and allow the slower moving patrol to see more of the action at the station. It also allowed personal interface with residents that could not be accomplished in a car. Many Navy and City municipal police departments have since copied it.

1993

February saw the new marina pier dedicated which more than doubled the number of slips available to boaters. The base was also running a quick scenario on what might happen if NAS Cecil Field was closed due to BRAC. Numerous options were looked at as impacts to NAS Jacksonville. Public Works employees were moving back into their building (902) in late February after a year long renovation. Unfortunately, the additions changed the historical aspects of the building as it was renovated. The average cost of a trouble call to have a minor repair made on station in 1993 was $67.80.

March 1 saw the Naval Supply Center change their name to Fleet and Industrial Supply Center, Jacksonville. The mission of supplying parts and goods to the fleet remained the same. March 12 and 13 saw the "Storm of the Century" hit the station. The main casualty was Building 13, the future home of the personnel who oversee construction of facilities on the station. It was being renovated, and a fire broke out on the night of 12 March. Due to the high winds, it was really never contained and the building was destroyed. What remained was demolished in May, all but the bottom four walls of the two-story structure. It was rebuilt in June. The building damage was very noticeable to all who traveled on Roosevelt Boulevard as it was right at the entrance to the station. March 13 saw the winds continue and forced the cancellation of the "Freebird" concert scheduled to occur at the station. The concert had to be rescheduled for the Civic Auditorium in Jacksonville for March 15. The winds of that weekend caused approximately $12,000 in damage to station roofs. The winds were also blowing badly for NAS Cecil Field as the list of activities to be considered for closure due to BRAC was announced, and NAS Cecil Field was listed. The NADEP at NAS Jacksonville was spared, however, for the time being.

April saw the Family Service Center become a station department under Commanding Officer, NAS Jacksonville. NADEP was also rolling out the first of 33 Greek A-7 Corsairs, with the first one completed March 25. Although the Navy rework program had ended, NADEP had won a contract to rework the aircraft the Navy was disposing of for the Greeks. On April 23, the budget cuts continued as Helicopter Squadron

NINE was disestablished. The Sea Griffins had been at the station since 1976 when they were reestablished as a squadron at the station.

Captain Cramer had the vision to start electronic mail (e-mail) for the station commencing May 4. At the station today, it is hard to envision how any business could be conducted without it. This one vision alone changed the business practices for the station immeasurably. Lieutenant Commander Kathryn Hire (now an astronaut) became the first woman to be assigned a Navy combat aircraft on May 23, when she was assigned to a P-3 of the reserve squadron VP-62. On May 25, the NADEP was added to the list of facilities to be considered for closure by BRAC. The reaction was concern, but not at a high level yet. The NADEP had 2,828 civilians and a $105.7 million payroll.

June saw the much-awaited next round of BRAC closings. NAS Jacksonville was concerned, but not worried. NAS Cecil Field and the NADEP, on the other hand, were worried! Employees of the NADEP made a bus trip to a hearing in Atlanta to voice their concern and present facts. NAS Cecil Field was announced for closure on June 26. June 13 saw reserve squadron VR-58 rename one of their C-9 aircraft. All of their planes are named after some city. The "City of New Orleans" was renamed the "City of Saint Augustine." June 27 saw a crash of a private aircraft in Concord, New Hampshire. Killed were Ron and Karen Shelly. They had performed a wingwalking act at both the 1990 and 1992 NAS Jacksonville air shows. Ron had earlier served at the station. Both were outstanding individuals who would be missed. June 26 was a lucky day for the NADEP as it was taken back off the list for closure consideration. Partrol Squadron 30 was busy training the first female operational combat pilot when Ensign Raquel Bini entered the squadron for six months of training.

July 13 saw the demolition of the old package store, Building 1884. It always seemed very odd to me from the time I entered the station that the package store (selling mainly alcohol and beer) was located on one side of the street and the Navy Alcohol Rehabilitation Center (helping those with a drinking problem) was located directly across the street! That oddity finally ended with the demolishing of the old store and the alcohol sales area moving to an area in the Navy

Exchange. July 14 saw a small fire start after a lightning strike on the roof of a supply warehouse, Building 164. The fire was noticed at 7:08 p.m. and was extinguished by the station fire department. Their quick action saved millions of dollars of stored supplies for the fleet. July 15 saw the latest on-station fatality to occur at least through 2000. A female member was walking along an SH-60 helicopter being towed, both attached to HS-1. She was carrying the wheel chocks and apparently decided to throw them into the open door of the helicopter instead of carrying them. Her foot got caught by the rolling wheel, and she was pinned and run-over by the helicopter. The accident was a horrible event for the station and one which occurred, oddly enough, during a mandated "safety-week" being held station wide. Captain Cramer, station Commanding Officer, was actually commencing a safety tour at building one when the accident happened. An investigation was conducted by the Helicopter Wing, and considerable action taken as a result of this terrible mishap. The day before a U.S. Customs HH-60 Blackhawk helicopter based at the station hit a utility wire and crashed in Georgia while on a drug interdiction mission, killing four. These accidents, combined with someone who had been drinking too much and fell off the hood of a car they decided to ride on Friday of that week near one of the clubs, made for the worst week in the stations safety history, instead of one that emphasized it. The station has not had a "safety week" since, and maybe won't as long as I remain safety officer!

The Family Service Center moved into their new facility, Building 876, in August. In retrospect, this move was done too far ahead of schedule by the Family Service Center director who wanted to dedicate the building with the Commanding Officer prior to her imminent departure. From the moment the personnel moved into these spaces, problems with the building arose. Today, FSC is looking for new spaces, as serious moisture problems causing mold growth on the walls continues. On August 18, the P2V Neptune static display was dedicated in the Aircraft Park on Yorktown Avenue. The plane was restored, again lead by Wayne White, with assistance from military personnel and the NADEP. Fred Church replicated the old VP-7 insignia on the side of the plane. This aircraft made a beautiful addition to the park. The VP-

7 markings were changed to the present VP-62 markings a short while later, as VP-62 "adopted" this aircraft to keep clean for display. This would be the last big ceremony for Captain Cramer, as he would have a change of command on August 20. His relief would be Captain Resavage, a helicopter pilot. In one of the most moving change of commands, I opened the official booklet only to discover Captain Cramer had included a crossword puzzle for people to do just in case they got bored with the ceremony! Such was the sense of humor and leadership the station lost when Captain Cramer left. Today, he is the Director of the American Red Cross in Jacksonville.

September saw Patrol Squadron 30 become the largest squadron in the Navy. The other training squadron, VP-31, had just been disestablished on August 26, making VP-30 the only P-3 training squadron.

The Human Resources Office became a separate command on October 1. NAS Cecil Field was also busy celebrating their 50th Anniversary with an air show extravaganza! It was especially well attended as the Citizens of Jacksonville came to see the station that was destined for closure. The last major event of 1993 saw the Naval Oceanographic Command change their name to the Naval Meteorological and Oceanographic Jacksonville Facility Atlantic.

1994

January started with the NADEP waiting to hear if they would be the new rework site for the F-14 Tomcat. The Tomcat workload was coming from the depot closing at Norfolk. The depot was busy hiring 100 workers as it would grow by 1,000 due to BRAC closure of three other depots. The first Tomcats arrived January 6, but not for rework. They would eventually be used for parts in the rework program. The NAS Jacksonville Plan of the Day distribution was stopped on January 19. It would be available on electronic mail in order to save paper. Also being disestablished on January 14 was patrol Squadron 49. They had a 50th Anniversary celebration and disestablishment ceremony at the same time!

February 8 saw the Resident Officer in Charge of Construction (ROICC) move into Building 13 near the main gate, after a renovation. The building,

which was destroyed by fire March the previous year, was actually improved to a better standard due to the fire.

The lawsuit filed by the base maintenance contractor, Apex, was settled in March. The suit ended up costing the Navy an additional $13 million. In awarding damages the panel who heard the case believed the base set out to sabotage the contract, thus the high monetary award. Joan Lunden of the Good Morning America show was visiting in March, going through water survival training. She was training for a flight to the USS Saratoga. The station was also gathering data for the next round of BRAC, to be completed in 1995. This would be the final round.

April saw the NADEP having a ceremony for the induction of the first of three P-3's to be worked for the Royal Thai navy and the disestablishment of the Navy Absentee Collection Unit. This unit would bring sailors who had gone absent without official leave (called UA in the Navy, for "unauthorized absence") back to the station. Their duties were transferred to NAS Security. April 2 saw an on-base fire in housing that destroyed the house. A three-year old was playing with matches, and started the fire. Damages were estimated at $30,000, but no one was hurt.

Friday May 13 saw a memorial service at the base chapel. Some security personnel had met to go to a concert in Orlando to volunteer for security assistance. Just as they pulled onto I-95, a car heading north went across the median and hit them airborne, head-on. Two of the officers were killed, and two seriously injured. The Scouts returned again to the station May 14, but with the crowds continuing to grow smaller. On May 19, the Defense Megacenter Jacksonville was established.

June 2 saw another name change for the Admiral's staff as Commander, Naval Aviation Activities Jacksonville became Commander, Naval Base Jacksonville. This change meant the command went from an operational control command to a non-operational control command. They now controlled the bases in the southeast region, Caribbean and Panama Canal. Also changing was Commander Helicopter Antisubmarine Wing ONE to Commander, Helicopter Wing Atlantic. They still controlled the helicopter squadrons based at the station. June 6-11 saw the railroad tracks removed on station. In a twist, the Construction battalion changed supervisors during the middle of the project. This led to them removing more track than was requested, stranding the locomotive inside the station with no way to hook up to the track along Roosevelt Boulevard. For the first time in the stations history there would be no more train service for anything. June 13 saw a sudden severe storm hit the station. Patrol Squadron 30 was most affected. A P-3 Orion was under the wash rack when the storm hit. The winds blew the plane through the wash rack, tearing off a section of the tail that blew into the parking lot at the Personnel Support Division. Squadron personnel were called and informed of the proper procedures for setting the brakes on a P-3! Approximately $2 million in damages were done to the aircraft. June 21 saw a meeting of Officers from Sea Control Wing Atlantic (who controlled the S-3 Viking Squadrons based at NAS Cecil Field, Commander Patrol Wing ELEVEN Officers (who control the P-3 Orion Squadrons at NAS Jacksonville), and the Commanding Officer of NAS Jacksonville. The purpose of the meeting was to explore the option of transferring the S-3 squadrons to NAS Jacksonville related to a BRAC decision to close Cecil Field. The S-3 community really wanted to stay in Jacksonville, and many local politicians supported them. This meeting would prove to be very fruitful, as a look at the ramp space at NAS Jacksonville shows today. The S-3 squadrons would be redirected to NAS Jacksonville in BRAC 1995.

July 7 saw Captain Resavage and NAS Jacksonville Supply Officer Commander Joe Clements breaking ground for a new five-story Bachelors Officers Quarters. The building was to be located on the St john's River right beside the 50 year old BOQ, Building 11. Since the station now had a train engine that was trapped on station with no track to go on, the 1940's engine was donated to the Florida Gulf Coast Railroad Museum in Tampa, Florida on July 19. The little engine had served the station well for over 50 years.

The 9th of August was the scene of a different kind-of meeting in the Commanding Officer's Office. On October 21, 1979, the Naval Investigative Service was burned out of their building on base. The individual who set fire to the building was eventually caught. Now, some 15 years later, the female who burned the building wanted to apologize to the Commanding Officer. Captain Resavage

heard her story, but was not in a position to forgive her since he was not the Commanding Officer in 1979 when the event happened. He did wish her well, however. After this, station personnel were apologizing to the Commanding Officer for running red lights, being late for meetings, etc. Also in June, the station maintenance budget was reduced to help defray the costs of the recent suit lost to Apex International, the former maintenance contractor. The USS Saratoga was decommissioned in a ceremony on August 20. Several NAS Jacksonville based squadrons had served on board the ship since it had been stationed at Mayport.

September 9 saw the crash of an SH-3 Sea King from HS-5 while operating in the Central Gulf, but there were no injuries this time, fortunately. In spite of the fact there over 15,000 personnel at the station on any given day, it is still a relatively tight community. The loss of any pilot is usually felt by all at the station and they were happy no one was lost in this mishap. September 17 saw another storm that did damage on station. This time, however, it was to an irreplaceable item. Along Mustin Road was the State of Florida Champion Oak tree. This was the tree Commander Stockton saved when the base was being built because it was so large and beautiful. On this day, a storm came through and it split right down the middle. One half fell over, and the other half remained, although leaning badly off balance. It had to be removed, thus ending the title NAS Jacksonville had for the Champion Oak tree in Florida. September 24-25 saw another air show at NAS Jacksonville. The shows continued to improve as the station fell into an every other year routine. Over 90,000 people attended the two-day event, again featuring the Blue Angels.

The NADEP was formally inducting the first EA-6B Prowler and the first F-14 Tomcat on October 3. Aircraft worked on previously were more or less to help the NADEP figure out how to work on the aircraft and get their new processes on track. These aircraft are a major part of the NADEP's workload today, along with the P-3 Orion. October 9 saw the freebird festival held at the station. Over 5,000 people came to the airfield to hear Lynyrd Skynrd and Charlie Daniels, plus some other local bands. Reserve squadron VR-58 had just received their fourth C-9B aircraft and were preparing to name it after another city in Florida.

The City of Starke, Florida had written an official request to have it named after their City, among others being considered.

December ended the year with the Chief of Naval Operations visiting the station, Admiral Boorda. The Fleet Aviation Specialized Operation Training Group also changed their name to Naval Aeronautical Medical Institute on December 21. This tenant of the station performs the water survival training all aircrew must go through. With the closing of NAS Cecil Field, they were also to start high-altitude training. The pool training is not well liked by the pilots, but necessary for survival if a crash occurs in water. Parachute landing in water is conducted, as well as dunker training. All pilots must be "dunked" into the pool in a simulated aircraft frame, which turns upside down and sinks to the bottom as it hits the water. In addition, the last dunking is while you are blindfolded! All tests are in your flightsuit with helmet. If you ever get a chance to tour the station, ask to see this evolution, if possible. It will give you a new understanding of the training performed at the station.

1995

On March 14, it was announced that a new role for the S-3 Vikings based at NAS Cecil Field would actually keep the squadrons in Jacksonville. The aircraft was being redefined as an all-around utility aircraft, and away from its traditional antisubmarine warfare mission. If the BRAC commission approved the shuffle, the planes and personnel assigned would not be leaving Jacksonville. And their personnel wanted to stay! Chief of Naval Operations Boorda was at the station again on March 22, speaking to station sailors. As he was speaking, the first steel was going up for Hangar 30, the future new home for Patrol Squadron 30.

April 10 saw the completion of the Building 27 renovation, which housed the Facilities and Environmental Department.. This building is the only structure remaining from the Camp Foster and Camp Johnston days, having been built in 1921. Another patrol squadron, VP-24, was disestablished on April 13,

April 28 saw fears rising at the NADEP yet again that BRAC 95 would announce the Depot would be closed. It had escaped the previous round of

BRAC, but the employee's were a little more uncertain this time around.

May 10 finally saw action on BRAC 1995. NADEP Jacksonville was actually on the initial list to be announced for closure. It was only through the actions of Congresswoman Tillie Fowler that the activity was removed from the list at the last second. On May 15, Congresswoman Fowler went to the NADEP and received thanks from all 3,000 NADEP employees during their lunch. May 12 saw Admiral Delaney return to the station. He had been given the assignment as Commander, Naval Base Jacksonville, and the senior commander to the station Commanding Officer.

June 1 saw the USS Saratoga finally towed away from its longtime home at Naval Station Mayport. The Defense Accounting Office at NAS Jacksonville was also disestablished. This office, formerly known as Regional Accounting and Distributing Center, handled civilian pay matters in addition to some other functions. Their job was transferred to an office in Pensacola. June 26 saw a ground breaking by Captain Resevage for a new Westside regional park, with an entrance to be located across from the main entrance to the station. The next day, NADEP was having a roll-out ceremony for their first completed A-7 Corsair for the Thai Navy.

July 5 saw the closing of the Chief's club on base. In a cost cutting move, the Chief's club was closed and the chiefs got a section of the enlisted club for their new location. The Navy Band, composed of two former bands with one relocated from Orlando and the other from Charlston, was also beginning practice at the station in the old Chief's club. Their future permanent home was moving toward a September finish. The band is actually assigned to the Admiral based at NAS Jacksonville, and not the Commanding Officer of the station. Also, "Old Reliable," an SH-3 Sea King helicopter, was leaving the station for Norfolk. With this last flight, the Search and Rescue (SAR) mission assigned to the station ended, after 45 years of service. SAR would now be rotated among the helicopter squadrons at Naval Station Mayport and the squadrons based at NAS Jacksonville.

August 25 saw Captain Robert Whitmire relieve Captain Resavage as Commanding Officer. A Naval Academy graduate, he had been in command of VP-5's Mad Foxes during Desert Storm. Four days later, Captain Whitmire was distributing his command philosophy to all employees. Listed were safety, leadership, environment, attitude, and his policy on sexual harassment, among some other items. He was also quick to stir things up at the station. He stated in a department head meeting he would strike fast, and he did. He noticed sloppy things like the official visitor spot for the Commanding Officer, was spelled wrong! He asked the security personnel to justify why they carried guns. If you came back with a good answer, he accepted it. If you didn't have a good answer, odds were it was going to be changed and in all honesty, probably should have been! He would become one of the most popular commanding officers the station has ever had. He had separate quarterly meetings with all civilians; enlisted and officers. The flow of information was excellent. Of even more notoriety were his Commanding Officer Hotline Calls, which he humorously answered in the JAX AIR NEWS. People grabbed the paper each week just to read that one column. A typical question was: "Skipper, do you know when the galley renovation will be completed?" Answer "Yes." Question: "Skipper, I was leaving work this week, driving down Birmingham Avenue and my truck was almost hit by a golf ball. The ball actually hit in the left lane, bounced over my truck, and nearly hit my windshield. Is there something that can be done?" Answer: "Yes. I recommend a closed stance, rotate your right hand a little, and slow down that back swing!" OK, one more! Caller: "Skipper, I'm calling to find out where I can call to report a UFO." Answer: "You can contact Executive Officer Captain Smith. He is from Roswell, New Mexico, and is old enough to remember the Roswell UFO landing in 1947. He can answer all your questions."

I personally can not say enough about the sense of humor Captain Whitmire exhibited during his tour, but I would also not want to meet him for Captain's mast. The same day as the change of command was on-going, a final inspection was being conducted for the new Navy band building. Their future building was actually what was the old station brig! On a sadder note for the same day, an F/A-18 Hornet, assigned to VFA-106 at NAS Cecil Field, crashed at the Whitehouse practice field. The pilot was killed instantly as the plane rolled and landed on the canopy upside down. This was a particularly hard crash for his fiancée, who was concerned about the safety of her future husband flying the Hornet. He had invited her out to witness the safety of the plane and she was in the Landing Signal Officer's shack, near the runway, as the plane crashed.

Admiral Delaney announced regionalization of base functions in September. On September 14, the station had a severe lightning storm. The airfield was hit some 37 times over a 200-foot area. Hangar 122 was hit doing some damage to the hangar. The odd thing lightning does when it hits one of the older hangars on base is to the resident bats. Lightning hits the hangar, and the voltage normally just passes through the metal structure and harmlessly feeds to ground. But sometimes during that process, the bats are stunned and they just drop from the top of the hangar! Most recover very quickly from being stunned, but every once in a while one just lays there dead! During this lightning event, a piece of the airfield was actually blasted away!

October 14 saw the dedication of the Navy Band building. This was the third time the station had a band located here. There had been a 20-year gap since the last band left, however. The band assigned was actually composed of members from Navy Band Orlando and Navy Band Charleston, as both bands were looking for a home due to BRAC closure of their respective bases.

November 14-19 saw the civilian employees furloughed as the Congress and the president struggled over the budget. Employees declared "essential" were required to continue to work. All of the Naval Aviation Depot employees were included in this category. Most of the other station civilian employees, however, were furloughed. (I particularly enjoyed the break as I said "good-bye" to my wife every morning who had to go to the NADEP to work, as I was declared "nonessential" and didn't do too much to argue it! In fact, I actually named another office employee as essential that could not afford the service break. In retrospect this was a good move, as we all eventually got paid whether we were there or not for the furlough period, infuriating my wife even more!)

December saw the gate guards removed from the station gates. In an effort to reduce costs and open-up the base more, the guards were removed at the direction of Commander, U.S. Atlantic Fleet. Captain Whitmire was actually reluctant to remove the gate guards. All of the message traffic coming into the station showed a need for *increased security* and not the opposite. He hesitated until he was called by Commander, U.S. Atlantic Fleet who ordered him to remove the gate guards. The problems in their removal started almost immediately as personnel were neglecting to register their cars and civilians not stationed at NAS Jacksonville were coming in wanting to use the restricted base facilities. But on the positive side, I have never seen the traffic flow smoother into the station during morning rush hour! Most station employees, and particularly the residents in housing, however, missed the secure feeling the gate guards provided for the station and their work environments.

1996

The year 1996 was a very busy one for the station and the station's history. Not only did a large number of events occur, but also the station was preparing for the 50th Anniversary of the Blue Angels with an air show at the station where they were formed. Initial plans were also being made to invite back R.M. "Butch" Voris, the first Blue Angels flight leader. Butch graciously accepted our invitation to return to NAS Jacksonville, a place he had not seen since 1947.

January 25 saw a visit from the Honorable John H. Dalton, Secretary of the Navy. He gave a speech to all at the station auditorium while at the station.

February 7 saw three events taking place. The first was initial tree planting was started in a new park on station, to be called "Patriots' Grove." The park was the idea of Admiral Delaney, who was now back at the station as Commander, Naval Base Jacksonville. The 79 trees planted would be to honor the Navy personnel who received the Medal of Honor since the start of WWII. The second event to take place on February 7 saw another C9B of squadron VR-58 being named after a city. Although the City of Starke, Florida had written the command asking for the plane to be named after them, the City of Orlando was selected. The Executive Officer of VR-58 was Commander George Westwood. (Note: He would become Commanding Officer of the squadron in May and later become an Emergency Preparedness Liaison Officer working with me in 1998. There have been few "characters" I have met

at the station over the years, but George would definitely be one of them!) The dedication would take place at the Orlando airport, in a rainstorm. The third event saw Jaguars quarterback Mark Brunell taking the controls of a P-3 with squadron VP-45.

February 10-19 saw a U-2 spy plane operating out of the station. The aircraft was using the station as a home-base as it flew exercises with the USS Kennedy. It was a strange sight for the station personnel. This aircraft, unfortunately, would only last a few more months, as it would crash in California killing the crew on a later flight. February 16 saw the Barnett Bank on station robbed. Captain Whitmire, who had removed the gate guards the previous December, was being requested to replace them due to the robbery. Approximately $1600 was taken. The robbery also showed a flaw in the security at the bank. As the robbery was taking place, a silent alarm was triggered. The alarm went to Barnett's operations center in Jacksonville. The operations center, in turn, would call back to the bank to verify the alarm, then they would call the police, who would call the base security department. Base security would then send personnel to close the gates to the station, while starting a search for the robber. By the time all this took place, the robber was long gone! Changes immediately made were to install a silent alarm to base security. February 19 saw construction start on a new U.S. Customs hangar at the station. Customs is a little know group at the station for the most part. Their new hangar would be a big benefit to Customs who desperately needed it, but it would be a loss of ramp space for parking aircraft at the station for the Air Operations personnel.

February 21 saw Captain Rice of the NADEP calling an all hands meeting. Things were not going well for the NADEP's bottom line. The NADEP was projected to loose $30 million for the fiscal year. Two aircraft recently reworked at the facility had to be returned for landing gear failures, and four P-3's were being returned due to faulty undercoating causing advanced corrosion. These aircraft would have to be restripped and painted, at depot expense. Captain Rice told all of the supervisors they had until the end of the year to turn a profit, or the NADEP would probably be gone in five years! NADEP in the meantime was also having problems testing their first Tomcat. A previous test on March 23

burned the tiles in the test house. This had to be repaired before further tests could be conducted on the A-7 Corsair and the F/A-18 Hornets. They had committed to completing six of the Tomcats and needed a way to test them. The City was not about to give a noise waiver, so they got innovative. They tested the first aircraft by going onto the runway and asking the tower permission to take-off. That would allow them to go to full military power on the engine. At the same time, they would have technicians checking the engines to the jet that had no intention of going anywhere! The tower personnel realized what was happening, and told the NADEP personnel not to pull that stunt again! For the second test (on a Sunday evening in February) they towed the Tomcat to NAS Cecil Field. Aircraft are routinely towed between NAS Jacksonville and NAS Cecil Field and have been for decades. When they got to the main gate at NAS Cecil Field, they discovered the wings were just slightly larger than the gate. The Executive Officer of the NADEP was called (Captain Kosiek) around 3:30 in the morning and asked if they should return to NADEP, or could they tear down a section of fence and have the NADEP pay to repair it. He said to tear a section of fence and NADEP would have it rebuilt, but this time with a larger opening! And to add to all of these concerns, privatization concerns were starting to hit the depot workers. How much work could be given to private contractors was being discussed.

March 14 saw NADEP finally testing an F-14 in the modified hush house at the station.

April 1 saw the first nuclear carrier dock at Naval Station Mayport, as the USS John C. Stennis came for a short visit. April 9 saw a brig escape, the first one ever recorded at the station. The brig personnel go across Allegheny Road in the mornings to get exercise. Upon return to the brig, approximately 6:00 a.m., one hid in the bushes. Missing on a head-count, all personnel were thought accounted for. Strangely enough, at 6:30 a.m., the escapee was given an "outstanding" on his bunk inspection! The escapee then scaled a fence where he headed to Orange Park and stole a car. He then used a phone to call his wife at Mayport to tell her he was coming and get some clothes ready! Since he was in the Brig for beating his wife, she called Naval Criminal Investigative Service (NCIS) and told the story, who called back the brig (7:45

a.m.) The brig, thinking they saw the prisoner, told the NIS personnel he was in the brig. The escapee again called his wife to tell her he was still on the way. She again called NCIS, who again called the brig. When the brig personnel double-checked the individual this time, they discovered their error! Security was called at approximately 8:00 a.m.. Station gates were closed and cars were checked. When security sent an officer to the brig to get a picture of the escapee to aid in their search, they discovered he had been gone approximately 2 hours, so they reopened the gates and stopped their search, knowing he was long gone! The escapee was finally caught scaling the fence at Naval Station Mayport, after a car chase along Mayport road at speeds of 100 mph. His car was stopped when it ran into the back of a People's Gas truck. The escapee then ran and two shots were fired at him! The final result was Captain Whitmire had to fire someone for the first time - the Brig Officer!

April 15 saw security start issuing the first magnetic badges. Personnel with these badges, when actuated, could enter through gates and checkpoints, even if not manned. On Thursday, April 18, a dinner was held honoring U. S. Navy Medal of Honor Recipients at the Officers Club. This was one of the finest events I have ever attended at the station. Three Medal of Honor Recipients were present for the dinner and the next day's dedication ceremony of Patriots' Park. They were LTJG Thomas J. Hudner (Korean War), BM1 James E. Williams (Viet Nam) and HM2 Donald E. Ballard (Viet Nam). Mr. Hudner was the one who was the most surprised by the events. He had received his Medal of Honor as a fighter pilot of squadron VF-32 at Chosin Reservoir in Korea on December 4, 1950. He had crashed his plane to try and rescue an injured fighter pilot that was shot down. He had packed the fuselage of the plane with snow to keep flames away, and requested helicopter assistance. He was unable to extract the injured pilot, however, due to his leg being crushed against the cockpit area because of his crash, and he eventually bled to death. The pilot he tried to save was none other than Jesse L. Brown, the first African-American to receive the wings of a Naval Aviator, and who had received those wings in 1947 at NAS Jacksonville! The next day saw the dedication of the park, with the honorable Charles Bennett as the honored guest speaker. The park is planted using historic trees and is worth a visit by anyone in Jacksonville.

An event kept totally out of the press and even away from most station personnel was Speaker of the House Newt Gingrich's visit the Naval Aviation Depot on Friday, May 17. Accompanying him were U.S. representatives Tillie Fowler and Cliff Stearns, along with Mayor Delaney. They received a quick overview of the Depot's mission and the few personnel invited to be a part of the brief were very impressed with the intelligent questions asked by the House Speaker. Just four days later, station personnel would be at a memorial service for Chief of Naval Operations (CNO) Jeremy Michael Boorda. Boorda had committed suicide on May 16. His suicide had a big impact on a number of personnel at the station, some of whom had known him personally. He had been an extremely popular CNO with the enlisted ranks. Another death happened at the station on May 15. Jay Beasley, Mr. P-3, passed away while attending a conference at the station. The hangar where the patrol aircraft are today is named after him. May 18 saw Scout World again at the station. The highlight was Lee Greenwood performing at the event. The crowd was down to around 4,000, but an evening fireworks show was added for the first time. May 28 saw an interesting accident report pass my desk. Seems a security officer was driving in front of the chapel when an 11-foot alligator attacked the front bumper of his car! The teeth marks were very noticeable. They managed to chase him back into Casa Linda Lake on station.

June 1 saw one Navy truck scratched at the station. The railroad tracks were being removed at the Birmingham Gate. As the contractor truck carrying the dirt lifted up it's load to dump it, the entire truck fell over sideways (as the load shifted) and smashed flat the Navy pick-up belonging to the Contracting Officer watching the job. It is assumed Vallencourt Paving's insurance carrier was happy to provide the Navy with a new truck! June saw a rash of bomb threats hit the station in the wake of the Oklahoma City bombing. The rash of false calls particularly affected the patrol squadrons in Hangar 1000. Personnel from the Naval Investigative Service would eventually place an agent in the squadron posing as an enlisted man, which would befriend the personnel making the calls. Once he actually saw them place another false call, six personnel were arrested. But this did not occur until August 14, making for a miserable summer for the squadron Commanding Officers, having to vacate their spaces on a regular basis. June 27 saw NADEP roll out the first A-7 Corsair for the Royal Thai Navy and June 30 saw a new auto emissions policy for the station. All cars would have to show proof of an emissions test, whether you lived in Duval County or not! This meant Clay County residents, who did not have emissions testing requirements, would now need a test to register their car to drive on-base. This requirement was actually directed by the Florida Department of Environmental Protection, and not driven by Captain Whitmire who was taking the heat from the station drivers.

July 3 saw the Navy Campus for Achievement move from their dungy offices in Building 8 to a consolidated location in Building 110. This move would allow for further expansion and better service for all sailors who wanted to take advantage of educational opportunities offered by Navy Campus. July 9 saw VP-30 given control of their new hangar; to be called Hangar 30 (changed from Hangar 1001). On July 10, the first P-3's entered as they were placed there for protection from approaching hurricane Bertha! July 12 saw the dedication of the Riverview Bachelor Officer Quarters. This facility was a design-build project, which meant the project could be designed and changes made or items added as the project progressed. This facility is one of the finest places to stay when visiting the station, if not one of the finest in the Navy. Captain Whitmire invited the entire staff to stay the first night, so any problems could be found prior to the first guests arriving. At a cost of $7.9 million, the five story, 111-room facility was like a 5-star hotel! Included were nine VIP rooms on the fifth floor, with a river view. My wife got to see her first manatees the next morning, playing right in front of the new facility. July 8 saw an announcement that NAS Jacksonville had won the Secretary of the Navy's Award for Achievement in Safety Ashore for the fourth time in eight years. (They must not have heard about our truck and alligator incidents!) July 31 saw the experiment with no gate guards end. The guards were back in place, tightening security back to where it had been before. The traffic started to back up in the mornings on Roosevelt Boulevard again, causing additional complaints.

A contractor employee was hit and killed crossing the street in the early morning hours of August 9. Oddly enough, he was hit by a medical corpsman assigned to naval Hospital Jacksonville, who was very upset by the accident. The contractor employee had run across the road in the middle of a block right in front of the car, wearing black clothes and where there was no lighting. Organizational changes were taking place on August 12 for the station's departments. Captain Whitmire had seen his department head meetings grow to a full room. Each person in the room also had to tell the Commanding Officer what was going on, or worse yet, pass on some problem. He was leaving the room with more tasks than he came with! So the departments were reduced in number, as well as were the special assistants. The former 31 people attending the meetings were reduced to 9. This format has basically lasted until today, but the number has grown back to approximately 17.

On August 20 the NAS Jacksonville "home page" went on line. This source of information remains today, and is an excellent way to get information about the station (. VP-30 finally moved all planes and equipment out of Hangar 1000 to their new hangar during August 20-22. August 29 saw a public meeting with the Citizens of Jacksonville to discuss the movement of the S-3 Vikings at NAS Cecil Field to NAS Jacksonville. There were complaints of noise that would accompany these planes, and thus opposition to the plan. But Captain Whitmire had a brilliant plan. As the discussions on noise were being registered, he had arranged for an S-3 Viking to fly over the school where the meeting was being held. So quiet was the jet, the citizens in the audience did not even notice. He stopped the meeting and asked if anyone just heard that noise. No one did, and he then explained an S-3 Viking had just flown overhead, and that would be the extent of the noise they would hear with the planes coming to NAS Jacksonville!

September 9 saw the dedication of the new Aircraft Acoustical Enclosure at the NADEP. This "hush house" was desperately needed to commence testing of the F-14 Tomcats. The hush house's equipment was actually relocated from the Naval Weapons Industrial Reserve Plant in Calverton, NY. September 17 saw the Westside regional park dedicated and September 28 saw VFA-203 become the first

squadron to leave Cecil Field due to BRAC.

October 2 saw the new CNO, Admiral Johnson, visit the station. An "Admiral's call" was held at the Chapel, with personnel allowed to ask the new CNO questions. On October 5 a farewell dinner was held for the station's XO, Captain Bob Coonan. He was very popular with the station personnel, and he got the traditional "roast." Every Commanding Officer (and most Executive Officers) gets a dinner with the staff that turns into a roast of the poor individual leaving. Even a special beer was made for Captain Coonan on this evening's event! Two days later, all station personnel got 59 minutes of "Admin Time" to go home early to secure their houses for another approaching hurricane. On October 10, the present F/A-18 Hornet was put on static display at the main gate. It had been a dream of mine to have a Blue Angel back at the main gate, showing all that the team was formed here. We had looked hard (and are still looking) for the teams original plane, an F6F Hellcat. With none to be found, I started looking for an F/A-18. Captain Kosiek, NADEP Commanding Officer, was most excited about the idea and finally called me one day to say he had a plane! This was not just not an old stricken F/A-18 painted in Blue Angels markings, but a real Blue Angel plane that had been recently assigned to the team. After stripping out the Avionics and engine, the plane was given to the station, where it remains on display today. Hangar 30 was officially dedicated on August 18, with VP-30 already operating out of the new $24 million hangar. On October 21, the final SH-3 Sea King assigned to an operational squadron flew from the station. Reserve squadron HS-75 would still fly the SH-3 Sea King until December 1999, however. On October 23, Captain Whitmire formally put the new NAS Jacksonville organization into effect. The 31 former departments and special assistants were reduced to 15 on the station organizational chart.

The Blue Angels 50th Anniversary air show was held the weekend of 26-27 October. Of all the air shows the station has held commencing in 1990, this one was the finest ever held. The crowd was by far the largest of any in the 1990's; the Blue Angels flew to perfection knowing this was the 50th Anniversary at the station where they were formed; and the crowd loved having the first Blue Angel, "Butch" Voris present to sign autographs. Butch

has since been a close friend of mine, as we trade e-mails and jokes on a weekly basis. He thoroughly enjoyed his stay in Jacksonville and has since told all who would listen no air show has ever surpassed this air show in class and making him fell welcome! A dinner with staff personnel was held with Butch the evening before. Being Butch and his wife Thea's escort, I drove their rental car to Whitey's fish camp for the dinner. Upon leaving, I was going a little too fast in the rental car in the construction area right along the fish camp, and got pulled over. Here I was with the First Blue Angel, as the station safety officer, pulled over by one of Clay County's finest! As he started to tell me what I had done wrong and how much it was going to cost (under a new law that just went into effect doubling the fine for speeding in a construction area), I said "Let me introduce you to Butch Voris, the first Blue Angel!" The Officer said a few more words, and then stopped and said "Did you say Butch Voris?" I said "Yes, would you like his autograph instead of mine on that ticket?" With that, he talked to Butch and let me go with a warning!! Butch had hoped to have a final flight with the Blue Angels. Captain Greg Wooldridge, the Blue Angels Commanding Officer, even had a special flight suit made for Butch. But the medical officer stopped his flight. Butch could go on a flight but there would be no maneuvers. But flying with the Blues in his mind meant letting everything go! Since that was not possible, he did not fly. It was a very hard moment for me to personally watch. The man who had formed the team, could no longer fly with them, and you could tell it hurt Butch very much! How would it have looked for him to fly with the team at NAS Jacksonville, on their 50th Anniversary and have something happen, though? But I think Butch still would have gone, given the opportunity!

November 8 saw the Healthcare Support Office dedicated at the station. This was a $1.3 million renovation of building H-2083. Keeping the sailor's health in mind, the Wellness Center had a ground breaking on November 13.

The year ended with two events occurring on December 16. The first one saw the Commanding Officer and his staff move out of building 1, so a renovation of that facility could commence. The Commanding Officer and Executive Officer went to two old trailer's for the project duration. The rest of the staff scattered throughout the base as space permitted. The Admiral's staff was also relocated to temporary spaces. NADEP also rolled out the last T-2 Buckeye training jet to finish that workload. Year 1996, busy as it was, finally came to a close!

1997

The year 1997 will probably be remembered as the year squadrons moved to NAS Jacksonville. Not since 1973, when the helicopter squadrons arrived, had the station seen so many squadrons being reassigned here. After another visit to the station by Secretary of the Navy John Dalton which occurred on January 29, February 4 saw the first squadron from NAS Cecil Field assigned to the station, administratively at least. VS-22 was apparently the first one assigned, but this never came to pass in actuality. On the same day, an S-3 Viking assigned to NAS Cecil Field crashed in the Gulf, killing all four on board.

Part of the problem facing Captain Whitmire was finding a space for the S-3 community. A contractor was hired who came up with a cost estimate to make all the improvements necessary for the move to work, estimated at $17 million. The Navy Audit Service reduced this cost figure to $12 million and the figure was reduced again by Commander in Chief, U.S. Atlantic Fleet to $8 million. This was the final number he had to work with.

When Captain Whitmire arrived for his tour in 1995, he asked what was being done to transition the S-3 community to NAS Jacksonville. "We have a plan in place, skipper" he was told. Upon looking into it, he discovered a young lieutenant assigned to the Air Operations department had developed what he considered an invalid plan, but beautifully colored! (It has long been a theory of mine that if you use a lot of "show," it will normally fly through or win that award!) It was going around everywhere as "the plan." He stopped it, and started to develop a new plan. A deal was made with reserve squadron VP-62 to move them out of their hangar (114) and into the patrol hangar, Hangar 1000. VP-62 personnel actually wanted to move there, as it made them closer to the other operational patrol squadrons. The operational P-3 community, however, was actually calling Captain Whitmire a "traitor" for suggesting a reserve squadron move next to the operational squadrons. It made perfect

sense to him, to have a P-3 squadron by the P-3 squadrons! His friendship with Admiral Hall, Chief of Naval Reserves, also helped. As part of the deal, the reserve community would throw in $1million for hangar improvements, to be used in improving Hangar 113 for NAS Cecil Field squadron VQ-6. This would also help move other scarce dollars to make the plan work for all the additional hangar improvements needed to move all of Cecil's S-3 Viking community. The most noteworthy thing about the plan developed by Captain Whitmire, however, was that all improvements were completed *ahead of schedule* and *under budget*! The transition was virtually seamless to the S-3 community, who themselves were on a tight move schedule. Hangar space became available and the new Bachelor Enlisted Quarters were finished just as needed.

Times remained tight for the station also. The Human Resources Office learned on February 18 they would be downsized and consolidated at Bay St. Louis (Gulfport), Mississippi. Their staff would be reduced from 120 to 30. The station staff in the meantime was meeting at the Florida Yacht Club for two days of cost cutting measures and to find ways to re-engineer the station. Numerous ideas were proposed as Captain Whitmire took notes.

March 3 saw another crash resulting in loss of life to station personnel as an SH-60 Seahawk helicopter assigned to HS-3 crashed in the Atlantic off of the coast of North Carolina killing all four on board.

April 25 saw an oak tree fall on the Commanding Officers on-base residence, damaging his house. On a humorous note, a station security officer knocked on Captain Whitmire's door two days later, on his birthday, to inform him a tree had fallen on his porch!

Patrol squadron VP-45 was moving from Hangar 1000 to the hangar VP-62 used to call home in May, so renovations could be made in phases to Hangar 1000. The phasing plan was necessary so the hangar could still be used while modifications were being made to accomodate for the S-3 community. Scout World was held 17-18 May, with a crowd of approximately 5,000. Although larger than the previous year, the event was still small compared to past years events. Naval Hospital Jacksonville settled a discrimination lawsuit also on May 14, with a settlement of $77,000. Normally these discrimination claims are filed against

the Navy and handled in house. The number of claims filed that are being worked at any given time is not advertised, but the money paid out by the Navy is approximately $25 million annually. Usually it is cheaper for the Navy to just pay to stop a case, rather than fight it. This is because the Navy has to pay for all future appeals, win or lose, and the appeal process seems to be endless for the person filing a claim. Patriots Park was re-dedicated on May 22, with Mayor Delaney as the principle speaker. Just prior to the ceremony, the Barnett Bank was robbed and the base secured to look for the robber. This time the gate closing was immediate, but the suspect was never found. It was not what the station officials wanted the Mayor to see. As he entered the station, cars leaving the base were being searched!

On May 27, Admiral Delaney and Captain Whitmire dedicated the Wellness Center. Then, at approximately 4:30 a storm hit the station. It was called a microburst and totally destroyed Building 136 on the station. In addition, numerous roofs, trees and street signs were damaged. The storm was particularly traumatic for the wife of the CO of the Naval Aviation Depot. She was on Yorktown Avenue and saw the building starting to blow apart in her review mirror. She called her husband on her cell phone as her rear car window shattered. He told her to drive right up to the front doors of the NADEP, which she did. It took three people to push open the doors to let her in as the winds blew! Initial estimated damage was $5-8 million; but the station received no additional funding to cover the costs. May 29 saw Helicopter Antisubmarine Squadron One's (HS-1) last flight prior to disestablishment. Helicopter #164103 left for a short flight at 0842 with the squadron Commanding Officer at the controls.

A change for the telephone system took place on June 1 as the prefix for most station numbers changed from 772 to 542. This caused confusion for the rest of the year because it would have cost extra funds the station did not have to place a message when you called the old number to notify you of the new number. So people calling only received a message "That number is no longer in service." With the hundreds of numbers that were affected, the cost was too much to bear. On June 19, the disestablishment ceremony was held for HS-1. The oldest helicopter squadron in the Navy was now gone! This squadron trained pilots to fly helicopters prior to going to an operational squadron. Either HS-1 (located in Jacksonville) or HS-10 (located in San Diego) which also trained helicopter pilots had to go. The Jacksonville squadron was picked. The Pelican's Perch Child Development Center was closed on June 30, as child care operations were consolidated at one location.

July 11 saw NADEP loading A-7 Corsairs on a barge for future disposal at sea. The NADEP was finally getting out of the A-7 Corsair business for good, and the remaining old planes would be used as an artificial reef. The last A-7 Corsair to be completed for foreign military sales was completed later in the month.

Building dedications were the order of the day throughout the summer as the Family Housing staff moved into their newly renovated offices and U.S. Customs moved into their new hangar both on July 18; the Allegheny Road housing project was dedicated on September 12; the new $23 million BEQ was dedicated on September 17, and; the galley finally reopened after a $300,000 renovation on October 1.

The S-3 community and Sea Control Wing Atlantic were officially welcomed to their new home in a wet ceremony at Hangar 1000 on October 28. The commanding officer's S-3 Viking of VS-24 was the first to pull up to the hangar. This move meant approximately 50 additional aircraft would now be assigned to the station, involving six squadrons. The other squadrons would move as they came off of deployments. Helicopter Antisubmarine Squadron Three (HS-3) returned from deployment the next day, and went to Hangar 124 as their new home. Hangar 122, where they had previously been located, was given to the Naval Aviation Depot for their expansion. Vice President Al Gore made a quick stop at the station on October 31.

The Blue Angels performed at the last air show to be held at NAS Cecil Field on November 1-2. Although the weather was perfect for an air show, the final Cecil Field crowd was small in comparison to past events. Building demolitions continued throughout the rest of the month as the old Navy exchange building was demolished, along with two other smaller buildings. Also, more VS squadrons were arriving as VS-30 arrived November 20, VS-31 arrived November 24 and VS-22 arrived December 12. The Sea Control Wing that controls the S-3 Viking Squadrons completed their move December 16, the same day the Barnett Bank closed after operating on the station for 53 years.

The year closed out with the dedication of the new air terminal. The old terminal had outgrown it's capacity, and Bill Myers had an idea to modify two old buildings, located side-by-side, into a new terminal. Although the project went well beyond his estimate, the facility in operation today was sorely needed and is a welcome addition if you are waiting for an aircraft. Captain Whitmire also allowed me to design and install some display cases to show some of the aircraft models of planes that have operated at the station.

1998

The base was well into transitioning the S-3 community by January. Only two more squadrons were due from Cecil Field. One, VS-32, would come from their present deployment and the other, VQ-6, would commence transferring. VQ-6 flew the ES-3, an electronic reconnaissance version of the S-3 Viking. In all, approximately 2,500 personnel and 48 aircraft finally moved to the station. January 8 saw construction halted at the old Chief's Club, Building 789. The building was being renovated into new offices for the Personnel Support Detachment (PSD). PSD is where all sailors check into the base, receive their military identification cards, if needed, and where records are kept for a lot of the station-based personnel. Their old offices had outgrown their usefulness and new spaces were badly needed, but a problem with asbestos was discovered. This immediately stopped demolition and extra money had to be found to remove the discovered asbestos. As of March 2000, PSD still had not moved into their new facility, and final construction was not scheduled to be completed until the summer of 2000.

February 4 saw groundbreaking for the new $1.6 million S-3 operational trainer to be built at the corner of Yorktown and Ajax. A second groundbreaking followed on February 26 for an addition to the water survival pool, completing a new $2.9 million Aviation Physiology Training Facility.

March 9 saw the post office finally move from it's home of over 56 years. The building it was in was beyond repair, and they desperately needed more room. The post office moved into the vacant former Pelican's Perch child care facility, which also had plenty of parking for patrons. March 23 finally saw VQ-6 move into their new spaces in Hangar 113, completing the moves of all S-3 Viking squadrons from NAS Cecil Field. The last squadron to arrive, VS-32, arrived on March 31 straight from being on a six-month deployment. The station's static display park was also growing in March. I had asked Captain Whitmire if he would support moving the static display aircraft located at NAS Cecil Field to NAS Jacksonville. I thought it would be a good idea to keep the aviation history in Jacksonville. (That and the fact I really like the aircraft static displays!) He agreed, so we wrote to the National Museum of Naval Aviation and asked for permission. We received all of the former Cecil Field aircraft except for three. Of those three, one went to the Museum, one went to NAS Oceana, and one went to a carrier for display. NAS Jacksonville received the rest, except for one, which no one wanted. The sole remaining static display at NAS Cecil Field today is the A-7 Corsair in front of their main entrance. This particular Corsair was actually the first operational Corsair to land on an aircraft carrier. We already had an A-7 coming to NAS Jacksonville for a future display, and did not need two. A-7 Corsairs were also being dumped, and were actually rather easy to obtain, if a museum needed one. So where is this static going that had such a significant place in naval aviation history? To the Don Garlitz's Drag Racing Museum along I-75 in Florida, at a future date!! The first of the Jacksonville additions, an S-3 Viking, went on display on March 24.

April 8 saw the change of command ceremony for Captain Turcotte to relieve Captain Whitmire. The ceremony was held in Hangar 30. On the Friday night before the ceremony, however, was the traditional "roast." I was fortunate enough to be asked to be the Master of Ceremonies and roast we did! Captain Whitmire took it all in stride, as one of the second largest groups I had ever seen came to have one last dinner with the "skipper." Captain Turcotte arrived as the first "jet" skipper the base had seen since the 1960's. The station Commanding Officer has to be an aviator, per Navy regulations. Since the station only had two main communities from the 1970's until the Sea Control Wing moved here from NAS Cecil Field, the Commanding Officer was either a patrol or helicopter community pilot.

Captain Turcotte was an S-3 Viking pilot. He did something no other Commanding Officer has ever done at the station since I have been there, that being inviting his entire staff over to his house the evening of his change of command. The word was "Come hungry and have fun!" The second former NAS Cecil Field static display was added on April 21 when a TBM Avenger from WWII was put on display, after a new paint job. Following right behind that one was an S2F Tracker, added to the park on May 5. The tracker was actually coming home. There had been two trackers located in the woods near the fire burn pits for over 20 years. When NAS Cecil Field was looking to start their static display park, I told them about the planes. They towed them to Cecil and made one good plane out of two. Now the plane was back at the station it had just temporarily left! April 30 saw the station library close, in a cost cutting move. All of the books, except a few navy books Captain Whitmire allowed me to "save" for historical purposes, went to the disposal yard for sale to the public. The closing caused an outcry from the public and even Congresswoman Fowler. But it had been identified in the meetings held the previous year as something the station could no longer afford in the wake of seriously reduced budgets. Budget concerns were also reaching a critical time. A message on April 21 announced station personnel could no longer make a service call to have maintenance items fixed. Included in this was even fixing broken air conditioning! Self-help was the new way to go.

May saw Admiral Delaney leave the Navy as he retired and a new Commander was assigned to the Naval Base. The ceremony for the change of command was one of the grandest the station had ever seen, with no less than two separate fly-by's of naval aircraft! May also saw the Morale, Welfare and Recreation headquarters personnel move from their home (Building 620) to another old building, 584. The move was made so that building and the former library building beside it could be demolished. A member of station-based HS-7 was killed in a training exercise on May 28, in Fallon, Nevada. His death was caused due to his falling out of the helicopter during a repelling event. May 29 saw the Subway shop open on station, as fast food establishments increased their visibility at the station.

June was a very interesting month! Playboy magazine came out and a station-based Navy flier was baring all!

This was definitely a confusing message for the post-tailhook Naval Aviation community! The female aviator could have faced disciplinary action for violations of Navy uniform regulations, or lack thereof! After a meeting with Admiral Delaney, she left the Navy, which she wished to do anyway. The first week in June saw the station assisting in fighting the Florida wildfires. Not only was the station assisting, they were directly supporting NAS Cecil Field who had fires of their own to contend with. All of their F/A-18 Hornets were temporarily located to NAS Jacksonville for safety. I had called Captain Turcotte and asked if he would allow the station to be designated as a Base Support Installation for use by the Federal Emergency Management Agency (FEMA). He immediately said yes, and the move was on to rush fire-fighting equipment and personnel through the station. Before the operation was over, 18 flights including C-5A's, C-130's, C-141's and even one 747 landed and off-loaded equipment. Approximately one million tons of cargo moved through the station.

Building demolition continued in July as Building 585 (located at the intersection of Enterprise and Gillis) came down 6-10 July and the old post office building, 920, came down July 15. By the end of July, the NAS Jacksonville supply department was moved from their home to temporary spaces while their building was renovated. They would also not return there after the renovation, as the spaces would be occupied by a new "regional" financial group, actually belonging to Commander, Naval Base Jacksonville, the senior command located at the station. There were 10 stations under this command, and all lost their resources management departments to make up this new staff. The financial group would handle all of the money for all stations in the region, including NAS Jacksonville's funds.

August started out with an accidental discharge of 900,000 gallons of wastewater that eventually flowed into the St. John's River over a 34-hour period. August 18 saw a Viking from VS-24 having landing problems at the station. The antiskid brakes failed, and the aircraft slid off of the end of the wet runway, after turning 160 degrees. The only damage was to runway lighting, fortunately. August 14 saw the jet trainers arrive at NAS Jacksonville. The squadron of TA-4J and T-45 Goshawk jets came from Corpus Christi, Texas.

Student pilots were here to conduct their first carrier landings on their way to earning their pilot wings. The jets were louder than most normal traffic heard around the station, making for some noise complaints. One early test for Captain Turcotte was a request for the jets to conduct touch and go landings at the station, instead of Whitehouse Field. The student pilots had to make five such landings prior to doing their first carrier landing. Deciding the 25 minutes of noise was worth it for the students, he approved the landings. As I watched the aircraft come and go in one continuous loop, I never thought it would be the last time one of the students would see the station, as he later crashed at sea while conducting his carrier qualifications. This would also not be the last crash of a plane in August, as a T-34 training plane assigned to NAS Cecil Field crashed in the woods near Baldwin on August 28. The plane was flying low and hit a power line, killing the pilot. The plane was available to pilots assigned to NAS Cecil Field to use for practice. The pilot killed was no novice flyer, as he was the safety officer for the F/A-18 squadron's senior command, the Strike Fighter Wing. August 20 saw an increased security posture for the station, as Threat Condition Alpha was instituted. The immediate impact for this posture was that cars would not allowed to be parked within 40 feet of any building. This eliminated a lot of parking at the station!

Demolitions continued in September as Building 620 (old station library) was bull-dozed along with the building right beside it, 621. Building 621 had been the Morale, Welfare and Recreation Department headquarters building. FEMA was again looking at NAS Jacksonville in September as the base was designated a mobilization site for approaching Hurricane Georges'. Semitrucks started rolling in off-loading equipment for use in supporting hurricane relief operations in Florida. Fortunately, it was not needed in this area, and it left as quickly as it came.

October 16 saw the last change of command take place at NAS Cecil Field as Captain Ken Cech relieved Captain Frank Bossio. Captain Cech had just left the Admiral's staff at Commander, Naval Base Jacksonville as the Chief of Staff. October 21 saw another tree added to Patriots' Grove, as Bob Ingram had just been awarded the Medal of Honor and a tree was placed in the Grove after him. The total number of trees in the Grove honoring Medal of Honor recipients was now 80. He also had an incredible sense of humor!! The U.S. Customs personnel finally dedicated their hangar October 23, probably to tie-in to the station's air show events, which went from the 23-25. Once again the Blue Angels were the main attraction.

Completion of the renovation of Building One was finally completed in early December, and on December 7 the staffs of NAS Jacksonville and the Commander, Naval Base Jacksonville (Admiral's staff) commenced moving back into their spaces. This move allowed the S-3 Viking senior command, Commander Sea Control Wing, to leave their temporary spaces in Hangar 1000 to go to their permanent spaces in Building 850, where the Admiral's staff had been. December 16 saw an A-7 Corsair return back to the station. In a difficult operation for me that took about six months to complete, the A-7 was returned from Jacksonville University where it had been taken from NAS Jacksonville some 5 years earlier. Jacksonville University wanted the plane on their campus and cut a deal with the National Museum of Naval Aviation. Now that they were getting a new football team, they needed the space where the jet sat and no longer wanted it. The problem was finding a set of slings so the jet could be airlifted back to NAS Jacksonville, since the Navy had gotten rid of all of these with the phasing out of this jet. The NMNA also did not approve of air lifts to transport aircraft, since they had recently had one fall during a lift. So, they were not told! The station Aircraft Intermediate Maintenance Officer finally found a set of slings and even had his personnel assist in rigging the jet for transport. Having been based with VA-46 at NAS Cecil Field, he was excited about repainting the plane at NAS Jacksonville and getting it on display at the station in 1999. I told him "If you can get the plane here, you can paint it any way you want!" (With the Commanding Officers approval of course!)

1999

January 13-15 saw all of the Commanding Officers of the Admiral's region (including Captain Turcotte at NAS Jacksonville) attending a conference on station. These meetings ended up defining what the regionalization process would finally entail. Although initially called Assistant Chiefs of Staff to the Admiral for program areas, their titles were changed to program managers at a later date. The region was now defined. Captain Turcotte, in addition to being Commanding Officer at NAS Jacksonville, would also be the program manager for all now nine stations under the Admiral for air operations and for Morale, Welfare and Recreation programs. Other station commanding officers were also given responsibilities, with the commanding officer at Naval Station Mayport given port operations and the commanding officer at Naval Submarine Base Kings Bay given security and fire responsibilities. This concept continues to be defined today, and probably will be for some time to come in the future.

The Naval Aviation Depot was having a change of command January 29, as Captain O'Neil was relieving Captain Kosiek. Captain Kosiek was leaving as one of the nicest commanding officers the NADEP had ever had! Captain O'Neil got his first taste of what it was like to be in charge at his change of command reception. He was pulled off to the side and notified that a forklift moving an F-14 Tomcat, recently reworked and destined to be returned to the fleet, had hit the plane while moving it away from the ceremony where it was on display for the crowd. Damages were minor, however, at $48,000.

The Admiral's staff was changing their name again in February, as they were now called Commander, Navy Region Southeast. This was a welcome change as the previous name, Commander, Naval Base Jacksonville, actually was being confused with Commanding Officer, NAS Jacksonville. February 3 saw a VFA-81 F/A-18 go off of the end of the runway, dropping it's ordnance! It was inert (no explosive) and no damage was done, except to the pilot's nerves! February 22 saw the NADEP induct an SH-60 helicopter for rework. This was the first helicopter to be seen at the facility in over 30 years. The A-7 Corsair that had been recently relocated from Jacksonville University was placed on static display in the aircraft park on February 26.

March 20 saw NAS Jacksonville hold the first Special Olympics at the station. This event had been held at NAS Cecil Field, and the station decided to host the event with Cecil's closing.

April saw the station's old power plant building demolished, Building 650. Also, on April 25, the P-3 squadrons' senior command, Patrol Wing Eleven, changed their name to Commander, Patrol and Reconnaissance Wing Eleven with the addition of their six aircraft.

Scout World returned May 14-16, now held in Hangar 30. The crowds were starting to build back up, after years of low turnout.

June 1 saw the dedication of the new weapons school for Sea Control Wing Atlantic, Building 852, after a $4 million project. The Naval Hospital lost some personnel to an auto accident in Haiti on June 4. Two personnel were killed and 11 injured in the accident. The new golf course expansion took off also in June as three buildings were demolished and tree removal started to add additional nine holes. A new parking lot opened also in June, but was not used! NADEP personnel had gotten used to parking in the field south of their engine shops. Not until Captain Turcotte closed this area and forced them to use the new lot would it be filled to capacity.

The last squadron to leave NAS Cecil Field, VFA-137, flew away July 14. With this squadron leaving, the closure of the station could be accelerated. Two more static display aircraft, the last to leave Cecil, also were moved to NAS Jacksonville on July 22. An F-8 Crusader and an A-4 Skyhawk were towed to the station. They are parked on the station ramp today, awaiting construction of their new pads at the static Display Park. This should be accomplished by the summer of 2000. July 25 and 26 saw two F/A-18 fighter squadrons (VFA-82 and VFA-86) assigned to NAS Jacksonville. The squadrons basically had no home! They returned from deployment unable to go back to NAS Cecil Field, and their new home, NAS Oceana, was not ready for them yet. So the squadrons were assigned to NAS Jacksonville. It was also during this period NAS Jacksonville had more planes on their flightline than at any time since the 1950's. The trainers from Corpus Christi again arrived, bring 56 aircraft. The F/A-18 squadrons added another 25. Some large transports had come to the station, and most of the S-3 Viking and P-3 squadrons seemed to be in town, all at once. The station was packed to capacity! Poor Bill Myers, airfield manager, and his staff were playing musical chairs trying to fit all of the aircraft into parking spots, while still maintaining clearance for airfield safety.

On July 27, the trees were being removed from a block surrounded by Yorktown, Saratoga, Gillis and Keily. A future concert was planned for this area, and the tree harvesting coincided just perfectly with plans.

August 3 saw the McDonalds restaurant on station robbed. With the closure of the Barnett Bank, guess the robbers needed a new place to hit! The robber's car was found in the contractor's lay down area, but no robber was ever caught. August 10 saw squadron VQ-6 send their last ES-3A to Arizona for storage. The squadron itself was disestablished on August 26. The duties assigned to this squadron were given to all of the other S-3 Viking squadrons. On August 27, F/A-18 squadron VFA-82 had their change of command ceremony in Hangar 114. This would be a first for NAS Jacksonville, having this fighter squadron ceremony at the station.

September 14 saw Hurricane Floyd approaching the coast of Jacksonville. All of the stations planes that could fly (approximately 100), were sent away. Those that could not were hangared. Personnel were sent home starting at noon, and the scheduled POW/MIA ceremony was cancelled. A small staff element, led by Captain Turcotte, stayed in the emergency operations command center. It wasn't until the weather personnel gave the 2300 (11:00 p.m.) briefing that things started to ease. The next day saw a deserted base, with Captain Turcotte already planning on how he could get a scheduled Friday evening concert back on track! The concert, headlined by the Goo Goo Dolls, Fastball and Sugar Ray, was a complete sellout, with 12,500 tickets sold (the maximum allowed.) It was also an overwhelming success, bringing some funds to the base's MWR program.

The F/A-18 squadrons assigned to the station, finally started flying out to their new home September 22-23. I hated to see them go, as they could have fit easily into the hangars at NAS

Jacksonville. The only problem was with their noise. With their leaving, the city of Jacksonville was without a fighter squadron for the first time since 1943. September 27 saw U.S. Customs announce a new squadron of P-3's would be based at the station. This move would allow the squadron to conduct surveillance on drug runners, in addition to providing another chip on the table for not closing the station in a future round of BRAC! September 30 saw the final actions to close NAS Cecil Field. Lieutenant "BT" Smith and I went to NAS Cecil Field to take down their sign and keep it for storage. We also witnessed the final lowering of the flag, conducted at 11:37 a.m. The flag was given to the last Commanding Officer, Captain Ken Cech. With that action, NAS Cecil Field was no more. Only a sign on the way out signaled what had happened, as it read "The fat lady has sung." The closure of Cecil was a great loss to Naval Aviation. Although no Navy Officer I talked to could admit it in public, all privately stated what a terrible mistake had been made by the navy in closing the station. Although the station may be developed into future commercial ventures that could benefit Jacksonville, the loss of the airfield to the navy is what may turn out to be the biggest loss of all. The station had little congestion, like Oceana. Since the F/A-18's have moved to Oceana, the noise complaints have been constant for that area in Virginia. (The move there may yet prove to be a wrong one, and I would keep an eye on future developments in regard to another future placement for these F/A-18 squadrons!)

October 7 saw Governor Bush hold a meeting at the station. Governor Bush was taking seriously another round of BRAC, and he was determined not to

let any of the Florida's military be lost without a fight! The meeting was a planning session on making sure Florida was as prepared as possible to defend any further military base closings. October 28 saw Blue Angel #7 crash in Valdosta, Georgia, killing two pilots. The Blue Angels were scheduled to fly at NAS Jacksonville the following weekend, but their appearance had to be cancelled due to this accident. The engines were immediately shipped to the NADEP for mishap investigation.

NAS Jacksonville's air show was held the following weekend (November 5-7) , to considerably reduced crowds from shows of the past. The show was only attended by approximately 80,000 visitors, but a new night air show that Friday evening was added and was a big hit! It will be expanded for the November 2000 show and moved to Saturday night. Station-based S-3 squadron VS-32 lost two pilots on November 14, as an S-3 Viking crashed after launch off of the USS Kennedy. One of the pilots lost was the son of a former Blue Angels leader. This was kept quiet, since the recent crash of the Blue Angels plane, which killed two, did not need additional press tie-in and stories. The ceremony held in the Chapel was one of the finest ever conducted at the station as a tribute to lost flyers.

December 1 saw the last three SH-3 Sea King helicopters fly away from the station. Reserve squadron HS-75 was receiving new SH-60 helicopters, and was phasing out the old birds. Of the three helicopters to leave, one went to Patuxent River, Maryland, and the other two went to HC-85 in San Diego. On December 8, a DC-8 was accidentally taxied into the U.S. Customs hangar, tearing off a two-foot section of the wing. It was repaired, and the plane sent on its way.

The millennium was ending on yet another positive note for the station. After a decade of cuts, dis-establishments, reorganizations and new leaders, the station finished the 90s with the same number of squadrons calling the base home. The Naval Aviation Depot was larger than it had started the decade, and it's future still looked bright.

The new century I am sure will bring about still more uncertainty, but this station seems ready to handle anything thrown its way. NAS Jacksonville remains one of the most requested duty locations in the Navy, and its people continue to be the reason this station succeeds when others seem to falter. The future looks bright for many more years of "Service to the Fleet."

This train engine, shown passing the static display area near the main gate, performed duties from 1940 until finally disposed of in July 1994. (U.S. Navy)

1990 Squadron

Disestablishments

VP-56 Disestablished June 28, 1991

PATROL SQUADRON FIFTY-SIX HISTORY

Since its establishment in 1946, the Patrol Squadron FIFTY-SIX (VP-56) "Dragons" have maintained an outstanding record of Antisubmarine Warfare excellence and professionalism under the leadership of numerous skippers, changes of aircraft, homeports and names.

The squadron was first established under the Naval Air Reserve in July 1946 as Patrol Squadron 900. While flying the PV-2 "Harpoon" aircraft at NAS Anacostia, Washington, D.C. the squadron was redesignated VPML-71, then VP-661. In 1950, VP-661 was called to active duty and following its transition to the PBM "Mariner" joined the Atlantic Fleet at NAS Norfolk, Virginia. In March 1953, the squadron acquired its present designation as Patrol Squadron FIFTY-SIX, and transition to P-5M "Marlin" seaplanes.

Beginning in January 1961, VP-56 received its first P2V-7 "Neptune" and began the transition from seaplanes to landplanes while in competition for the coveted Battle Efficiency Pennant, which the squadron won in July 1961. In addition to the Battle "E" VP-56 was awarded the Captain Arnold J. Isbell trophy for Excellence in Air Antisubmarine Warfare.

In October 1962, VP-56 deployed five aircraft to U.S. Naval Air Station, Guantanamo Bay, Cuba. With the establishment of the Cuban Quarantine, the entire squadron received orders to Cuba. For its efforts the squadron

received a letter of commendation from Admiral H.P. Smith, Commander-in-Chief U.S. Atlantic Fleet and in September 1963, received a second Battle "E.

Coincident with its change of homeport to NAS Patuxent River, Maryland in 1967, the Dragons began transitioning to the P-3B "Orion." Preparations for fleet introduction to the P-3C "Orion" began in early summer 1969 and in September, VP-56 became the first operational squadron in the United States to fly the most advanced ASW platform of its kind in the world. The squadron changed homeports to NAS Jacksonville from NAS Patuxent River, Maryland on July 2, 1971.

Since then the Dragons have made numerous successful deployments to Sigonella, Sicily; Bermuda; Rota, Spain; Lajes, Azores; and Keflavik, Iceland, maintaining the highest standards of performance. In both at home and deployed operations, VP-56 has attained unequaled recognition for ASW operational proficiency. Specifically the Dragons made naval aviation history by being the first patrol squadron to receive two Sixth Fleet "Hook 'Em" trophies for ASW excellence in a single deployment, while operating out of Sigonella in 1982. The Dragons earned their third "Hook 'Em" trophy in as many years during the Rota/Lajes deployment of 1983.

The squadron again deployed to Sigonella in January 1986 and returned to NAS Jacksonville in July after completing one of the most successful MPA deployments ever recorded in the Mediterranean theatre. Through the deployment, VP-56 proved to be an invaluable asset to the Sixth Fleet carrier battle groups during operations in the vicinity of Libya, and received the Navy Unit Commendation, the Meritorious Unit Commendation, the Armed Forces Expeditionary Medal, and a Sixth Fleet "Hook 'Em" award for ASW excellence. In February 1987, VP-56 received its third Battle "E" and surpassed 23 years and 157,000 hours of accident free flying in July 1987 while deployed to Bermuda. During the most recent deployment, the Dragons served as the East Coast VP representative for UNITAS XXVII, operating safety and effectively from nine South American countries in a six month period, in addition to meeting the operational requirements in Bermuda.

The Dragons deployed once again to Sigonella, Sicily, from January to July 1989, completing another extremely successful deployment. Upon their

return, the Dragons commenced Update III transition. In April 1990, the Dragons earned the patrol Wing ELEVEN Retention Excellence Cup (2nd Quarter) and on July 21, 1990, the Dragons surpassed 26 years and over 174,000 hours of mishap-free flying.

The Dragons completed the first full deployment of the P-3C Update III aircraft in the Keflavik Sector on February 8, 1991 and received orders on February 11, 1991 to disestablish as a squadron. June 28, 1991, marked the end of an era of ASW excellence as the Dragon's retired their colors.

HS-17 Disestablished July 2, 1991

HELICOPTER ANTISUBMARINE SQUADRON SEVENTEEN HISTORY

Helicopter Antisubmarine Squadron SEVENTEEN (HS-17) was established under a blazing spring sky on April 4, 1984 with 71 initial plankowners. During the establishment ceremony, the squadron's crest was unveiled, proudly introducing "Neptune's Raiders" to Helicopter Antisubmarine Wing ONE at NAS Jacksonville.

HS-17, homeported for their entire existence at NAS Jacksonville, joined Carrier Air Wing Thirteen, serving most recently aboard the USS Coral Sea (CV43). They were also at their normal compliment of 26 officers and 160 enlisted personnel. The "Valkyries" provided close-in ASW support for the CV Battle Group. While embarked aboard the Coral Sea, HS-17 was unique in that it served as the sole organic ASW asset. Neptune's Raiders also provide search and rescue, plane guard, anti-ship missile defense, and logistics support, flying six SH-3H Sea King Helicopters.

The squadron's first deployment came fast and furious commencing in October 1985 and ending eight months later in May 1986. The squadron made Naval Aviation history by being the first to deploy an H-3 to the Black Sea. During March and April 1986, CVW-13 assets flew combat sorties against Libyan military targets as part of Operation Eldorado Canyon, which was in response to terrorist acts against United States citizens. HS-17 was awarded two Navy Unit Commendations, the navy Expeditionary Medal and the Armed Forces Expeditionary Medal for its support of the three Carrier battle groups during the freedom of navigation and Eldorado Canyon operations.

HS-17 again deployed to the Mediterranean with Carrier Air Wing Thirteen aboard USS Coral Sea (CV-43) from September 1987 to March 1988. While deployed, the squadron participated in a number of joint allied exercises. Neptune's Raiders returned to a rousing welcome on March 29, 1988.

Throughout the remainder of 1988, the Raiders honed their ASW tactical skills through aggressive use of submarine services at the Atlantic Underwater test and Evaluation Center (AUTEC). A highlight for HS-17 was winning the "Commander's Trophy" during HSWING One's ASW "Professional Week" and being awarded CVW-13's "Excellence" award as the top squadron for the USS Coral Sea work-up cycle. HS-17 completed it's third cruise to the Mediterranean aboard the USS Coral Sea. This was the final cruise for this exceptional carrier, nicknamed the "ageless Warrior". While deployed, the squadron defended the carrier's inner zone from submarine threats and provided strike rescue support during contingency operations off Beirut, Lebanon. The Raider's received their second carrier Air Wing Thirteen "Excellence" award in recognition of their sustained superior performance throughout the deployment. 1989 and 1990 were capstone years for HS-17 in that the squadron swept all competitive maintenance and operational awards during this period, including back-to-back HSWING One maintenance Trophies, the Captain Arnold J. Isbell award for ASW excellence, the COMNAVAIRLANT Battle Efficiency award the Admiral John S. "Jimmy" Thatch award as the top carrier based ASW squadron in the Navy. In addition, the squadron was awarded the Meritorious Unit Commendation for the 1989 Mediterranean deployment.

With the decommissioning of USS Coral Sea, HS-17 was left without a ship and soon after, without an airwing. The indeterminate squadron status since October 1989 did not deter the squadron from maintaining the highest state of readiness. The squadron completed carrier qualification detachments, cyclic flight operations and workups aboard the USS Theodore Roosevelt (CVN-71), USS Enterprise (CVN-65) and deployed in September 1990 on USS Abraham Lincoln (CVN-72) on her homeport change from Norfolk, Virginia to Alameda, California. In May, HS-17 completed her support to USS Kitty Hawk (CV-63) as she returned to active service from the Service Life Extension program (SLEP). The squadron had been nominated for a navy Unit Commendation Award (third) for their impressive accomplishments performed under demanding conditions while sustaining outstanding readiness.

HS-17 stood ready to meet current and future needs of Naval Aviation. The squadron's last Commanding Officer was CDR Stephen J. Bury, as they disestablished in an impressive ceremony on July 2, 1991.

HS-9 Disestablished April 30, 1993

HELICOPTER ANTISUBMARINE SQUADRON NINE HISTORY

The Navy's first all-weather capable helicopter squadron, HS-9 was established on June 1, 1956 at Naval Air Station Quonset Point, Rhode Island. During the 1960s the Sea Griffins participated in the quarantine of Cuba and aided in the evacuation of casualties from the USS Liberty, which was attacked in the Mediterranean during the 1967 Arab-Israeli conflict.

The squadron was disestablished in

October 1968, and called back to duty at NAS Jacksonville in June 4, 1976. The Sea Griffins joined Carrier Air Wing 8 aboard the USS Nimitz, where they participated in the filming of the movie "The Final Countdown."

HS-9 aircraft have operated from the decks of many ships, including 11 carriers from the historic HMS Ark Royal to the U.S. Navy's newest, the USS Theodore Roosevelt. The squadron has shown the flag in more than twenty countries, and operated from the Caribbean Sea to the Indian Ocean and from the waters north of the Arctic Circle to south of the Cape of Good Hope. During a recent deployment of the USS Roosevelt, the squadron set a record amount of submarine contact time.

VP-49 Disestablished January 14, 1994

PATROL SQUADRON FORTY-NINE HISTORY

Through the years, the Patrol Squadron Forty-Nine "World Famous" Woodpeckers have set the standard of excellence for Patrol Aviation with their aggressive performance both in the air and on the ground.

The squadron's roots trace back to World War II when on February 1, 1944, it was established as Patrol Bombing Squadron Nineteen at NAS Alameda. The Squadron saw WWII combat action while flying missions in the Pacific in the PBM-3D seaplane. In 1945, the transition to the Marlin PBM-5 Mariner began. Restationed to NAS Norfolk, Virginia in 1946, the squadron operated as Medium Sea Plane Squadron Nine (VPMS-9). It was here on September 1, 1958 that the squadron was designated Patrol Squadron Forty-Nine. The "Forty-Niners" moved east to the island of Bermuda on July 5, 1951 and it was here that the transition to the Martin P-5M Marlin occurred.

In the early 1960s the squadron returned to the United States and was stationed at NAS Patuxent River, Maryland. In 1963, a new era dawned as the squadron began operating its first land based airplane, the Lockheed P-3A Orion. This was followed in 1969 by transition to the computerized P-3C aircraft. Patrol Squadron Forty-Nine finally arrived at its present home base, NAS Jacksonville, Florida on January 31, 1972.

Patrol Squadron Forty-Nine's history is rich in events familiar even to those who are not acquainted with the VP community. The sacrifices of World War II, assisting with evaluation trials of the Nautilus, America's first nuclear powered submarine, actively involved in the blockade during the Cuban Missile Crisis, participating in recovery operations for projects Mercury, Gemini and Apollo manned spaceflight programs, flying Market Time and Yankee Station Patrols off the coast of Vietnam, and most recently, supporting the interception of the Achille Lauro highjackers during the 1985 Sigonella, Sicily deployment. In 1986, the Woodpeckers traveled to South America to participate in the Unitas XXVII joint exercise with the Navies and Air Forces of Peru, Chile, Uruguay and Brazil.

In 1989, VP-49 transitioned to the Navy's most sophisticated multi-mission maritime patrol platform, the P-3C Update III. Two years later, the squadron commenced integration of the "Beartrap" program. With these technological advances, the Woodpeckers continued record setting "on-top" performances against Russian new construction submarines while deployed to the Mediterranean and North Atlantic. The Woodpeckers final deployment to Keflavik, Iceland, was highlighted by unprecedented employment of the navy's "From the Sea" principles through cutting edge US Air force joint interoperability and ground breaking special warfare initiatives. Lastly, on December 28, 1993, the Woodpeckers completed their final operational sortie. This milestone marked 32 years and 214,000 hours of accident free flying. Clear testimony to the squadrons unfailing and total commitment to excellence.

Patrol Squadron Forty-Nine's sustained performance can be readily seen in the list of awards received: 5 Navy Unit Commendations, 7 Meritorious Unit Commendations, the Coast Guard , 3 CNO Safety Awards and 8 Hook 'Em awards. The last Commanding Officer was CDR Mark H. Anthony.

VP-24 Disestablished April 13, 1995

PATROL SQUADRON TWENTY-FOUR HISTORY

Patrol Squadron 24 was originally commissioned as VB-104 on April 10, 1943 at NAS Kanehoe, Hawaii. The squadron was cited twice by the president during WWII "For extraordinary heroism against the Japanese." After the war, the squadron was redesignated VP-104, called many places home and flew the P4Y-2. In July 1948, it was redesignated Patrol Squadron 24. In 1955, the squadron was redesignated FA-(HM)-13. This stood for Attack (Heavy Mining) Squadron 13. The squadron's mission was mine laying. In 1956, the squadron returned to its current designation of VP-24.

After World War II the squadron participated in the testing and development of the "Bat," the Navy's first air-to-surface guided missile. As a result, the nickname "Batman" has carried down through the years. The squadron's famous "Batgirl" insignia is the only authorized U.S. Naval squadron insignia displaying a member of the fair sex.

The squadron relocated to NAS Jacksonville from Patuxent River on October 30, 1972. The Batmen received numerous awards in 1983, including the "Golden Wrench" maintenance award given to the command with the best P-3 maintenance record in the Atlantic Fleet, and the Meritorious Unit Commendation for operational performance. In addition, VP-24 was nominated for the Battle "E"

Award, the CNO Safety Award, and a "Hook 'Em" award for 1989.

HS-1 Disestablished June 19, 1997

HELICOPTER ANTISUBMARINE SQUADRON ONE HISTORY

Established on October 3, 1951 at Naval Air Station Key West, Florida, HS-1 was the first helicopter squadron with an antisubmarine warfare (ASW) mission. It remains one of the oldest helicopter squadrons in the Navy. Through its operations HS-1 demonstrated the effectiveness of the helicopter as an ASW weapon. This led to the commissioning of 10 other HS squadrons.

The squadron initially flew the Piasecki HUP. In June 1960, HS-1 changed from an operational ASW squadron to a training squadron. Today, the squadron flies the Sikorsky SH-3G/H "Sea King" in excess of 6,000 hours per year, qualifying approximately 75 pilots, 100 ASW aircrewmen and 180 search and rescue aircrewmen annually. This squadron also trains nearly 350 maintenance personnel annually. Pilots and maintenance personnel from more than 11 different foreign countries have received training at HS-1.

VQ-6 Disestablished August 26, 1999

Fleet Air Reconnaissance Squadron Six (VQ-6) was initially established at NAS Cecil Field on August 8, 1991. They assumed the carrier-based reconnaissance role that was previously performed by the EA-38 Vikings.

The "Black Ravens" flew the ES-3A, a signal intelligence modification of the S-3 Viking anti-submarine aircraft, which replaced the EA-38, a veteran of over 40 years of fleet service.

VQ-6 transferred to NAS Jacksonville in August 1998, one of the last squadrons to leave Cecil Field. Almost from the time they got settled in to their new home, word came that they would be disestablished. They sent their last ES-3A Viking to Arizona for storage on August 11, 1999 and were disestablished on August 26, 1999.

Top: SH-60F Seahawk flies "downtown." This aircraft sports HS-1 markings. (U.S. Navy)

Middle: SH-3 Sea King of HS-1 over downtown Jacksonville. (U.S. Navy)

Bottom: Squadron VP-5, in front of their P-3 Orions. (U.S. Navy)

One wonders what Captain Mason, the station's first Commanding Officer, might think if he could revisit NAS Jacksonville today. Would it today have realized his dream of 60 years ago? Some of the base's roads and buildings would be as familiar today as they were when they were built in the early 1940s. But a majority of the buildings and roadways have changed dramatically. The actual missions, except for the mission of the Naval Aviation Depot and the Naval Hospital, have also changed since the base's inception. Even though the Naval Aviation Depot and Naval Hospital still serve the same basic purpose as when they were first formed- reworking aircraft and treating personnel, the technology of today would boggle the minds of personnel who worked in these facilities in the early 1940s. The overall mission and station motto of "service to the fleet" remains foremost, however. A view of NAS Jacksonville's squadrons and wings today shows how complicated this "city in a city" really is, and probably how little most people, including a majority of the station's personnel, really know about the 3778 acres called NAS Jacksonville.

Commander, Navy Region Southeast

Commander, Navy Region Southeast (CNRSE) is the senior Navy command on board NAS Jacksonville. The local admiral has administrative responsibilities that include evaluation of shore activities, standards of performance, security, disaster preparedness, joint services, financial management, public affairs, state and local government liaison and legal affairs. He is also the immediate superior in command of the commanding officer of Naval Air Station Jacksonville and the commanding officer's of all other Naval stations in the southeast United States and the Caribbean. This senior level command, located at NAS Jacksonville since 1942, has been represented by either a one or two-star billet since early inception. This command, however, has gone through a mind-boggling series of name changes and missions, establishments and disestablishments.

Initially, the command was called Commander, Naval Air Operational Training Command and was responsible for all aviation-related training in the southeast during WWII. In 1948, this was changed to Commander, Fleet Air Jacksonville and the command controlled all operations-related aviation assets in the Jacksonville area. In 1973, this name was changed to

Commander, Sea-Based Antisubmarine Wing One along with a short-lived Commander, Tactical Air Atlantic which was also added. Tactical Air Atlantic was moved back to Norfolk, Virginia, one year later. Next came Commander, Helicopter Wings Atlantic in October 1, 1986; then Commander, Naval Aviation Activities Jacksonville on October 1, 1992, followed by Commander, Naval Base Jacksonville effective June 2, 1994 and finally the latest and current name for this command (Commander, Navy Region Southeast) effective February 2, 1999.

In the late 1990's, a regionalization effort took place that resulted in functions normally assigned separately to each station, being combined into a "regional" effort. Financial management consolidation was the first example of this effort. Additional studies are currently being taken, all in an effort to reduce costs and become more streamlined. The command currently has 16 officers, 100 enlisted personnel and 125 civilians assigned and presently is the Immediate Superior in Charge for nine naval activities in the Southeastern United States, also including Guantanamo Bay, Cuba and Roosevelt Roads, Puerto Rico. Also attached to Commander, Navy Region Southeast are Navy Band Jacksonville and the Fleet Imaging Center. The command headquarters is located on the first deck of the main administration building at NAS Jacksonville, Building 1.

Helicopter Antisubmarine Wing, U. S. Atlantic Fleet

Helicopter Antisubmarine Wing, U. S. Atlantic Fleet (formerly Helicopter

Antisubmarine Wing ONE) was formed April 1, 1973, at NAS Quonset Point, Rhode Island. The initial composition of the Wing included five squadrons: Helicopter Antisubmarine Squadrons ONE, THREE, FIVE, SEVEN and ELEVEN. The Wing's mission is to train and support carrier based (CV/CVN) helicopter antisubmarine squadrons capable of conducting all weather, multi-sensor Antisubmarine Warfare (ASW), Combat Search and Rescue (CSAR), comprehensive over water search and rescue (SAR), and logistics support for detachments aboard aviation capable ships.

Following its commissioning, HSWINGLANT was directed to relocate to NAS Jacksonville on December 15, 1973, when NAS Quonset Point was closed. Concurrent with the move south, Wing squadrons continued regular deployments aboard Atlantic fleet aircraft carriers. In November 1973, Helicopter Combat Support Squadron TWO was assigned to HSWINGLANT.

Helicopter Antisubmarine Squadron SEVEN was temporarily reassigned to Helicopter Sea Control Wing ONE for Interim Sea Control Ship (ISCS) evaluation in February 1974, returning to the operational control of HSWINGLANT once complete. Helicopter Antisubmarine Squadron FIFTEEN was commissioned and assigned to the Helicopter Sea Control Wing in October 1971 and was officially reassigned to HSWINGLANT in November 1973. Helicopter Antisubmarine Squadron NINE was reestablished in September 1977, but was again disestablished in April 1993 as a result of force reduction. Helicopter Antisubmarine Squadron SEVENTEEN was commissioned in April 1984, and was disestablished in June 1991. Helicopter Antisubmarine Squadron ONE was disestablished in June 1997. HSWINGLANT presently consists of five operational squadrons (HS-3, HS-5, HS-7, HS-11 and HS-15), the Weapons Training Unit (WTU) and the Surface Rescue Swimmer School (SRSS).

HSWINGLANT squadrons consistently perform their missions in an exemplary manner. Primary among these missions is antisubmarine warfare conducted in close proximity to aircraft carriers and other high value support units. In the Mediterranean, HSWINGLANT squadrons have been involved in numerous ASW exercises and missions which have included Tunisian flood relief, support for U.S. Task Forces involved in the October 1973

Middle East crisis, air evacuation of American Nationals during the Cyprus conflict, sea/air recovery of commercial airline disaster victims in the Eastern Mediterranean and support for task forces involved in the 1976 Lebanon crisis and Grenada. Recent events include helicopter operations in support of hostage rescue attempts in the vicinity of Bosnia-Herzegovina and support for operations DESERT SHIELD, DESERT STORM, PROVIDE COMFORT, DENY FLIGHT and SOUTHERN WATCH. Helicopter Antisubmarine Squadron ONE forward deployed attachments to assist in the relief efforts following Hurricane Andrew in August 1992.

When not embarked, Wing squadrons are kept on SAR alert and have made numerous rescues in the Jacksonville area. Whenever there is a requirement for ASW protection, logistic support, or SAR, the men, women and helicopters of the squadrons in HSWINGLANT are trained, equipped and operationally ready to launch.

Currently, HSWINGLANT has over 1,000 men and women assigned. Whether flying from carrier decks as proud members of the Carrier Air Wings, from aviation capable ships, or from shore bases, HSWINGLANT squadrons will continue to provide professional ASW protection and perform critical support missions for all Atlantic Fleet units.

SIKORSKY SH-60F/HH-60H "SEAHAWK"

All CHSWL Helicopter Squadrons at NAS Jacksonville fly the twin-engine Sikorsky SH-60F/HH-60H helicopter. The HH-60H's primary mission is Combat Search and Rescue (CSAR) and Naval Special Warfare support. The SH-60F's primary mission is anti-submarine warfare (ASW) defense of the CV/CVN inner zone, which includes detection, classification, and destruction of hostile submarines. Secondary missions are CSAR, and Naval Special Warfare support. Additional missions performed by both aircraft are logistic support, vertical replenishment, anti-surface warfare and medical evacuation. The maximum speed of the helicopter is 180 knots. The helicopter has an endurance of 4.5 hours and a crew of 4 is normally on board. The SH-60F is equipped with the AN/AQS-13F dipping sonar and can deploy active and passive sonobuoys. Armament includes one M-60D/M-240 machine gun and up to three MK-46/MK-50

torpedoes, sonobuoy launcher and a stores jettison system. The armament for the HH-60H consists of two M-60D/M-240 machine guns or two GAU-17A miniguns, a stores jettison system and aircraft survivability equipment.

HELICOPTER ANTISUBMARINE SQUADRON THREE

Helicopter Antisubmarine Squadron THREE was established on June 18, 1952, at the Naval Air Facility, Elizabeth City, North Carolina. The Tridents commenced operations flying the Piasecki UH-25B helicopter, and later transitioned to the Sikorsky H-19 and SH-34 helicopters. HS-3 was also the first East Coast operational squadron to transition to the SH-3 Sea King and then the SH-60F Seahawk helicopters.

In 1962, HS-3 participated in the naval blockade of Cuba while deployed on the USS WASP (CVS 18). The Tridents started their active role in the Space Program on August 24, 1962, when CDR J. M. Wondergem picked up LCDR M. S. Carpenter from his Aurora 7 spacecraft and delivered him to the USS INTREPID. HS-3 won the Isbell Trophy for ASW excellence and the Battle "E" for both 1963 and 1964. The squadron also flew its 25,000th consecutive accident free hour in 1964.

In the early 1970s, the squadron participated in the relief of Tunisian flood victims, rescuing or relocating 630 people while transporting over 43,000 pounds of food and medical supplies. The Tridents were awarded a Meritorious Unit Commendation for their efforts. HS-3 won the Battle "E" in 1973 and 1978 as well as the Isbell Trophy in 1974 and 1978.

HS-3 was awarded two Navy Unit Commendations for operational accomplishments and outstanding maintenance efforts during the 1985-86 Mediterranean and Indian Ocean deployment onboard the USS SARATOGA (CV 60). The Tridents won three consecutive Battle "E" Awards for 1985, '86 and '87. They also won back the Isbell Trophy in 1986 and '87. In 1988, HS-3 forged new ground for the HS community when they supported the first six month deployment aboard a Spruance Class Destroyer, the USS HAYLER (DD 997).

In 1990, the Tridents enforced United Nations sanctions against Iraqi trade while deployed aboard USS SARATOGA in the Red Sea. HS-3 also flew the first ever actual Helicopter Visit, Board, Search and Seizure (HVBSS) with a Special Forces boarding team to "take down" a hostile merchant ship during Operation DESERT SHIELD/STORM. The squadron returned home in the spring of 1991 and commenced its transition to the SH-60F Seahawk.

In 1993, HS-3 deployed with SH-60s on the USS THEODORE ROOSEVELT (CVN-71) and was instrumental in providing Combat Search and Rescue (CSAR) forces in the Adriatic and Red Seas while supporting United Nations sanctions against Bosnia-Herzegovina and Iraq, respectively. The Tridents also teamed up with the USS ARLEIGH BURKE and tracked a Kilo class submarine that was enroute to Iran. The ROOSEVELT/Trident team received a Meritorious Unit Commendation for its dedicated efforts.

Preparations for the squadron's 1995 cruise we interrupted when the Tridents were called on to support Operation UPHOLD DEMOCRACY during the military intervention in Haiti. HS-3 provided the sole Navy Maritime SAR, CSAR and special operations support. The Tridents deployed one

week after the completion of UPHOLD DEMOCRACY and flew in support of Operation DENY FLIGHT over Bosnia-Herzegovina. On deployment, HS-3 also flew the first actual CSAR mission by an HS squadron since the Vietnam era to search for the crew of a downed French Mirage fighter.

The Tridents deployed onboard the USS JOHN F. KENNEDY during the summer of 1996 to the North Atlantic. The deployment was highlighted by the carrier's visit to Ireland. During its spring 1997 deployment on board the USS JOHN F. KENNEDY, the squadron participated in Operation DELIBERATE GUARD off the coast of Bosnia-Herzegovina and Operation SOUTHERN WATCH, the support of United Nations sanctions against Iraq.

In 1998 HS-3 assisted local efforts in locating and fighting brush fires that ravaged Northeast Florida. The Tridents also became the first operational squadron to utilize the FLIR/Hellfire system on the HH-60H and scored direct hits on all firings.

1999 saw the Tridents deployed aboard CVN-71 in March and headed straight into the first of two conflicts. During Operation NOBLE ANVIL, HS-3 supported combat operations against Serbia in Kosovo. Once cease-fire was agreed upon, the TR transited to the Arabian Gulf to support MIO and enforce no-fly zones over southern Iraq. In August, HS-3 conducted a successful HVBSS to a freighter violating UN sanctions. The operation seized $3.5 million in Iraqi contraband. During this demanding cruise, HS-3 lifted over 1,800,000 pounds of cargo and completed over 2000 small deck landings. HS-3 was awarded the CVW-8 Golden Wrench, the HSWINGLANT Maintenance Trophy (2 awards), and the Battle "E" for 1999.

HELICOPTER ANTISUBMARINE SQUADRON FIVE

Helicopter Antisubmarine Squadron FIVE was commissioned at Naval Air Station, Key West, Florida, on January 3, 1956. Since its origin, the squadron has expanded upon its original mission: denying the enemy effective use of their submarines. HS-5 today is a true workhorse squadron, not only conducting its original Anti-Submarine Warfare missions, but also conducting missions ranging from Combat Search and Rescue to logistics in support of the Battle Group.

From the beginning, HS-5 has been

a pioneer in helicopter Anti-Submarine Warfare (ASW). The squadron was the first to be fully equipped with 14 HSS-1 "Sea Horse" helicopters. In 1958, HS-5 received the first HSS-1N helicopter. Subsequent testing and evaluation proved that helicopters could be hovered at night over water without visual reference to the ocean, and that they were ready for around-the-clock, all-weather anti-submarine operations. It was due to this pioneering effort with the "dipping" (submersible) sonar equipped HSS-1N that HS-5 acquired the nickname NIGHTDIPPERS.

In 1959, HS-5 was assigned to Carrier Anti-Submarine Air Group FIVE-FOUR and changed its home port to Naval Air Station, Quonset Point, Rhode Island. While there, the squadron transitioned to the first twin jet-turbine helicopter, the SH-3A "Sea King," later upgrading to the SH-3D with more powerful engines and improved dipping sonar.

1972 marked the beginning of a new era for the NIGHTDIPPERS. HS-5 was not only permanently assigned to Carrier Air Wing SEVEN, but also changed home ports to its present home, Naval Air Station, Jacksonville, Florida. Since that time, the NIGHTDIPPERS have made extended deployments on USS INDEPENDENCE (CV 62), USS DWIGHT D. EISENHOWER (CVN 69), and USS GEORGE WASHINGTON (CVN 73). Highlights during this period include a 315 day deployment in the Indian Ocean with 157 consecutive days at sea in 1980, 3000 hours in support of the Multi-National Force in Beirut in 1983, the first ever six month detachment of a single SH-3H on the USS PETERSON (DD 969), and the support of Operations DENY FLIGHT and SOUTHERN WATCH.

1995 began a new chapter in NIGHTDIPPER history when they transitioned from the venerable SH-3H

"Sea King" to the state of the art SH-60F/HH-60H "Seahawk." In 1996, as a result of superb teamwork and a passionate commitment to excellence demonstrated during their first operational deployment with the new SH-60F/HH-60H, the NIGHTDIPPERS were awarded the CNO Safety "S", the Captain Arnold J. Isbell Award for ASW Excellence, the COMNAVAIRLANT Battle "E", the Thatch USW Excellence Award and the Commander Sixth Fleet "Hook'em" Award for Undersea Warfare Excellence. The NIGHTDIPPERS were also nominated for the Captain Arnold J. Isbell Trophy in recognition of superior USW performance. This operational achievement was based on maintenance excellence which included both the CVW-7 Golden Wrench and two consecutive Commander, Helicopter Anti-Submarine Wing, U.S. Atlantic Fleet Maintenance Awards.

In late February 1998, the NIGHTDIPPERS deployed with CVW-7 in USS JOHN C. STENNIS (CVN 74) on her maiden Voyage around the world from Norfolk, Virginia to San Diego, California. During the deployment, HS-5 spend over four months in the Arabian Gulf in support of Operation SOUTHERN WATCH and returned to Jacksonville, Florida in late August 1998.

The end of the century marked the completion of workups for HS-5 in preparation for their February 2000 cruise aboard USS DWIGHT D. EISENHOWER (CVN 69).

disestablished on May 31, 1966 and established again at Naval Air Station, Quonset Point, Rhode Island, on December 15, 1969. The official insignia was redesigned to a configuration similar to the design you see today.

During the 1970s, the Shamrocks of HS-7 deployed to a variety of locations, including Vietnam and the Mediterranean Sea. In 1973, HS-7 joined Carrier Air Wing THREE, changed homeports to Naval Air Station, Jacksonville, Florida, and transitioned to the SH-3H helicopter.

From 1981 to 1993, the Shamrocks of HS-7 deployed onboard the USS JOHN F. KENNEDY (CV 67), and were called upon to serve during Operations DESERT SHIELD, DESERT STORM and PROVIDE PROMISE. In December 1993, the Shamrocks and Carrier Air Wing THREE shifted to USS DWIGHT D. EISENHOWER (CVN 69) and in early 1994 became the first operational HS squadron to embark with women. HS-7 was called upon in September 1994 to support "IKE" off the coast of Haiti during Operation UPHOLD DEMOCRACY, and later supported the IKE/CVW-3 team flying in the Arabian Gulf during Operation SOUTHERN WATCH and in the Adriatic Sea during Operations DENY FLIGHT and PROVIDE PROMISE. Upon return from deployment in 1995, the Shamrocks transitioned to the Sikorsky SH-60F and HH-60H, and completed their first deployment flying these new aircraft aboard USS THEODORE ROOSEVELT (CVN 71) in May 1997.

Following the transition to the H-60 aircraft, the insignia design was updated to reflect the silhouette of today's aircraft. Green and white are the colors assigned to the Shamrocks of HS-7 and are prominent in the design. The seven stars across the top reflect the

seven stars in the "Big Dipper" constellation, which served as the original insignia for the squadron. The aircraft with its sonar, submarine and aircraft carrier are all graphically depicted. The lightening bolts converging on the submarine illustrate the destructive power of the SH-60F and HH-60H weapons system and the men and women who wield them in the defense of the United States.

HS-7 recently returned from a six month deployment on board USS ENTERPRISE (CVN 65) where we participated in Operations DESERT FOX and DELIBERATE FORGE. HS-7 will start preparing later this year to deploy on board USS HARRY S. TRUMAN (CVN 75) in late 2000.

HS-7 proudly answers to the call sign "Dusty Dog" and provides the Navy with the Fleet's finest carrier vital-zone ASW defense, Search and Rescue, logistics, anti-ship missile defense, and Combat Search and Rescue capabilities. In addition to aircraft carrier operations, the Shamrocks provide detachments to air capable ships and regularly conduct ASW training at the Atlantic Undersea Test and Evaluation Center, Andros Island, Bahamas, and Combat Search and Rescue (CSAR) and special warfare support training at NAS Fallon, Nevada.

Numerous awards have acknowledged HS-7's accomplishments. Recent awards include the 1996 CINCLANTFLT Golden Anchor Award; the Commander, Helicopter Anti-Submarine Wing U.S. Atlantic Fleet Maintenance Award for the first half of 1997; the 1997 Commander, Carrier Air Wing THREE "Golden Wrench" Award for JTG 97-1 deployment; the 1997 Helicopter Anti-Submarine Wing, U.S. Atlantic Fleet nomination for the Secretary of Defense Phoenix Award for Maintenance Excellence; and the Commander, Carrier Air Wing THREE "Golden Wrench" Award for JTG-99-1 Deployment. HS-7 is recognized as a leader in rotary wing aviation and as one of the Navy's finest squadrons.

HELICOPTER ANTISUBMARINE SQUADRON ELEVEN

Helicopter Anti-submarine Squadron ELEVEN was commissioned June 27, 1957, at Naval Air Station, Quonset Point, Rhode Island. It was from this homeport that the Dragonslayers earned their original radio call-sign of "Snowbound." The squadron remained at Quonset Point until October 17, 1973, when the

HELICOPTER ANTISUBMARINE SQUADRON SEVEN

Helicopter Anti-Submarine Squadron SEVEN was originally established in April 1956 at Naval Air Station, Norfolk, Virginia, for the mission of harbor defense. HS-7 was soon assigned the role of ASW support for the Fleet. The squadron was

Dragonslayers moved to their present home of NAS Jacksonville, Florida. In 1989, the squadron petitioned for and received permission to change the squadron radio call-sign to its current call-sign of "DRAGONSLAYER." The squadron's mission of Anti-submarine Warfare (ASW) has been its primary, but by no means its exclusive, domain over the years.

When first commissioned, HS-11 was flying the Sikorsky SH-34 helicopter. In 1962, the squadron transitioned to the Sikorsky twin engine SH-3A "Sea King," the first helicopter specifically designed for antisubmarine warfare. HS-11 subsequently transitioned to the SH-3D and SH-3H, which were improved versions of the SH-3A. The SH-3H was equipped with improved sensors that provided greater operational capability than previous ASW helicopters. In addition to the pilot and copilot, two aircrewmen were positioned in the cabin area to operate the aircraft's detection equipment and interpret incoming sensor data. The aircrewmen were also trained to perform Search and Rescue (SAR) operations and to conduct ship-to-ship personnel and cargo transfers.

HS-11 transitioned to the Sikorsky SH-60F and HH-60H "Seahawk" in 1994, the current aircraft in use today by the Dragonslayers. Through transitioning to the Seahawk helicopter, HS-11 was able to strengthen its role in the ASW arena as well as expand into new mission areas. The Dragonslayers now added such missions as Vertical Replenishment (VERTREP) and Naval Special Warfare (NSW). With the addition of the FLIR/Hellfire Missile system in 1999, the Dragonslayers introduced the HH-60H into the new mission of Anti-surface Warfare (ASUW).

The Dragonslayers' excellence in the field of ASW has been well recognized throughout the years. HS-11 has been the recipient of several honors, including numerous Captain Arnold Jay Isbell Trophies for ASW excellence. The Dragonslayers have also won several Admiral "Jimmy" Thatch trophies for being the Navy's best carrier based ASW squadron. HS-11 was awarded the Commander in Chief, U.S. Atlantic Fleet Golden Anchor Award in 1983, 1984, 1995, 1996 and 1998. HS-11 has also been the recipient of the Battle "E" on numerous occasions.

The squadron has played a vital role in recovery operations of the nation's astronauts. HS-11 helicopters have been seen on national television picking up such famed astronauts as White and McDivitt after splashdown of their Gemini IV spacecraft in 1965. The squadron was then on scene again later that same year for the recovery of Gemini VI and VII after their successful rendezvous. As the Nation watched on live television, an HS-11 helicopter hoisted astronauts Lovell and Aldrin from their capsule at the conclusion of the twelfth and last Gemini flight.

Over the years, HS-11 has rescued many downed aviators and transferred untold numbers of sick or injured personnel of various nations to medical facilities. HS-11, while embarked on USS WASP, rescued 14 crewmembers from the foundering Liberian tanker SS PEGASOS that was sinking in the Caribbean. On November 22, 1975, HS-11 played a major role following the collision at sea between USS JOHN F. KENNEDY (CV 67) and USS BELKNAP (CG 26). The squadron flew a total of 36.1 hours of SAR and Medical Evacuations (MEDEVAC) for injured personnel. On September 14, 1976, a collision occurred between USS JOHN F. KENNEDY and USS BORDELON. HS-11, demonstrating their superior professional skills, again flew MEDEVAC pickups in marginal weather while also making external hull damage reports and area searches for other survivors. Squadron aircrews rescued more than 80 injured personnel during these incidents in adverse weather and sea conditions. For its meritorious service aboard the USS JOHN F. KENNEDY, the squadron was awarded the Navy Unit Commendation.

HS-11's contributions to the development of new tactics and equipment have played a major role in carrier-based anti-submarine warfare.

HS-11 was the first Atlantic Fleet squadron to successfully refuel a helicopter in flight from a destroyer, the USS PETERSON (DD 969). This greatly enhanced mission capability by extending on-station time in the submarine danger zone. In 1971 the squadron successfully evaluated the multi-channel jezebel relay system and tested new magnetic anomaly detection tactic known as "SHOELACE."

In October of 1973, HS-11 changed homeport from NAS Quonset Point, Rhode Island, to NAS Jacksonville, Florida, falling under administrative control of Helicopter Anti-submarine Wing ONE. After conducting pre-deployment exercises on board USS SARATOGA (CV 60) in the spring of 1974, the squadron, under new operational control of Carrier Air Wing One (CVW-1), deployed onboard USS JOHN F. KENNEDY in the summer of 1975.

At the end of January 1976, HS-11 returned to Jacksonville after a seven month Mediterranean deployment, during which the squadron flew over 3000 hours. In July of that same year, HS-11 was cited for achieving five years of accident free flying. After pre-deployment exercises, HS-11 embarked aboard USS JOHN F. KENNEDY again, for a North Atlantic deployment from September to November. During this deployment, the Dragonslayers participated in two NATO exercises, TEAMWORK '76 and BONDED ITEM.

In January 1977, HS-11 began an extended deployment to the Mediterranean Sea embarked on the USS JOHN F. KENNEDY. While conducting ASW exercises off the coast of Crete on March 3, 1977, the squadron surpassed 20,000 hours accident free flying to mark a major milestone within the helicopter aviation community.

In November of 1977, the Dragonslayers were called upon to perform a shakedown cruise aboard the Navy's newest nuclear powered aircraft carrier, the USS DWIGHT EISENHOWER (CVN 69). In 1978, the squadron completed yet another cruise aboard the USS JOHN F. KENNEDY (CV 67) while remaining at the forefront of the helicopter ASW community. During 1979, HS-11 stood down for a brief period, and sent numerous detachments to the Atlantic Underwater Test and Evaluation Center (AUTEC) to perform ASW testing and training.

In March 1980, the Dragonslayers completed a very successful Mediterranean deployment while again embarked on the USS JOHN F. KENNEDY (CV 67). During the eight-month cruise, the squadron flew over 3000 hours in support of ASW, logistics and SAR missions. HS-11 was also the first HS squadron to deploy a helicopter detachment to a SPRUANCE Class Destroyer during special operations.

In 1984 the squadron performed six rescues of Naval Aviators who had ejected from their aircraft. The squadron was awarded the Battle "E" Award, Isbell Trophy, COMNAVAIRLANT's "Silver Anchor" Award and CINCLANTFLT's "Golden Anchor" Award for an unprecedented second consecutive time. During the 1984 deployment, squadron aircraft were the first HS aircraft to be outfitted with M-60 machine gun mounts to further expand the capabilities of the venerable SH-3H. Additionally, during their eight-month Mediterranean/Indian Ocean cruise, HS-11 was the first squadron to experiment with night vision devices. During that deployment the squadron was awarded a Navy Unit Commendation and Navy Expeditionary Medal for superior professional accomplishment.

During 1985, HS-11 once again was awarded the Isbell Trophy and completed five years and 16,500 hours accident-free flying. The squadron also was awarded the COMSIXTHFLT "Hook 'em" Award for ASW excellence. The Dragonslayers participated in Operation OCEAN SAFARI in the North Atlantic and was awarded a Meritorious Unit Commendation for its superb tactical performance in the Vest Fjord, Norway.

During HS-11's Mediterranean cruise in 1986, the squadron participated in Operation PRAIRIE FIRE in Libya and was awarded a Meritorious Unit Commendation and Armed Forces Expeditionary Medal.

In 1988, the Dragonslayers were the first rotary wing squadron to experiment with dropping live bombs to counter surfaced submarine threats. The squadron flew numerous sorties to the Pine Castle Target Range and evaluated the possibilities of a new aircraft configuration.

The Dragonslayers, as part of CVW-1, departed Norfolk on 27 December 1990 embarked aboard USS AMERICA (CV 66) as part of coalition forces in support of Operation DESERT SHIELD. Shortly after the USS AMERICA (CV 66) reached the Red Sea, Operation DESERT SHIELD became Operation DESERT STORM. During the

42 day conflict the Dragonslayers flew over 1600 hours in support of the U.S. led coalition forces. In February 1991 in the North Arabian Gulf, Dragonslayer 612 located and destroyed an Iraqi floating mine within five miles of USS AMERICA.

Within three months of their return from DESERT STORM, the Dragonslayers began an intense inter-deployment training cycle in preparation for a second deployment in 1991. This included a visit to New York City for Fleet Week where DESERT STORM veterans were honored for their service to our country. Leaving New York City, the Dragonslayers set sail aboard the AMERICA for NORTHSTAR 91 and FLEETEX 1-92.

During this period the Dragonslayers received national coverage after rescuing three sailors who were adrift in a life raft for ten days as a result of Hurricane Bob. The Dragonslayers were also responsible for rescuing a KA-6D crew who ejected after losing an engine during their catapult launch in Vest Fjord, Norway. The Dragonslayers began a six-month deployment on 2 December 1991 and rescued two down aviators during the initial Carrier Qualification (CQ) period. In February 1992, USS AMERICA (CV 66) returned to the North Arabian Gulf. HS-11 worked in conjunction with multinational forces to refine tactics developed during Operation DESERT STORM and Operation SOUTHERN WATCH. On 25 May 1992 the Dragonslayers and USS AMERICA transited the Suez Canal for the forth time in eighteen months, returning to NAS Jacksonville on 6 June 1992.

In September 1993, the Dragonslayers and CVW-1 deployed again aboard USS AMERICA in support of Operation PROVIDE PROMISE in the Adriatic Sea and PROVIDE COMFORT off Mogadishu, Somalia. This was the last deployment HS-11 made with the SH-3H Sea King. They returned from deployment in February 1994 and transitioned to the SH-60F and HH-60H Seahawk helicopters in April 1994. This year, the squadron was awarded another Battle "E" Award, the CVW-1 "Golden Wrench" Award and the HS Wing Maintenance Award.

The Dragonslayers deployed aboard the USS AMERICA for her final voyage to the Mediterranean Sea and North Arabian Gulf in support of Operation DENY FLIGHT and DECISIVE EDGE. Their performance throughout the deployment earned

them considerable recognition. Squadron awards included the Commander in Chief, U.S. Atlantic Fleet Golden Anchor, the Captain Arnold Jay Isbell Trophy, the Commander, Naval Air Forces, U.S. Atlantic Fleet Battle "E" Award, the Commander, Carrier Air Wing ONE Maintenance Excellence Award and a Navy Unit Commendation.

On October 3, 1997, the Dragonslayers began a six month deployment aboard USS GEORGE WASHINGTON (CVN 73). The squadron participated in Exercise BRIGHT STAR in Egypt with multi-national military forces, and operated in the Arabian Gulf in support of United Nations sanctions against Iraq. The squadron returned home to Jacksonville, FL on April 3, 1998. For its outstanding performance throughout the deployment, the squadron was awarded the Commander, Naval Air Force U.S. Atlantic Fleet Battle "E" Award, the Chief of Naval Operations Safety Award, the Commander, Helicopter Antisubmarine Wing U.S. Atlantic Fleet Maintenance Award, the Captain Arnold Jay Isbell Trophy and the Admiral J. S. Thatch award for ASW excellence.

From August to December 1998, the Dragonslayers detached multiple times onboard USS HARRY S. TRUMAN (CVN 75). HS-11 won the Commander, Helicopter Antisubmarine Wing U. S. Atlantic Fleet Maintenance Award for the first half of 1998 and the CINCLANTFLT "Golden Anchor" Award for education and retention excellence.

During COMPTUEX/JTFEX in June 1999, HS-11 fired a direct hit with a Hellfire missile on the retired destroyer USS LAWES. This was the first direct hit on an unaugmented war-ship sized target proving the FLIR/Hellfire was a viable Anti-surface weapon for the HS community.

On September 15, 1999, two HS-11 aircraft launched from USS JOHN F. KENNEDY into Hurricane Floyd to rescue nine people stranded in the storm. The crews received international attention for conducting the rescues in winds in excess of 50 knots and in seas of 30 feet.

On September 22, 1999, the Dragonslayers deployed to the Arabian Gulf onboard USS JOHN F. KENNEDY (CV 67). During this deployment, King Abdullha II, King of Jordan, piloted an HS-11 HH-60H helicopter from Aqaba, Jordan to the USS JOHN F. KENNEDY underway in the Red Sea. The event went into the record books as the first U. S. Navy aircraft flown by a king of Jordan. In addition, the squadron participated in Exercise Bright Star in Egypt with multi-national forces and Operation SOUTHERN WATCH in the Arabian Gulf. The squadron returned home from its six-month deployment in March 2000.

HELICOPTER ANTISUBMARINE SQUADRON FIFTEEN

Helicopter Antisubmarine Squadron FIFTEEN was commissioned to fly the SH-3 Sea King on October 29, 1971 at Naval Air Station, Lakehurst, New Jersey.

During the next two and a half years, the Red Lions deployed aboard the USS GUAM as part of the Sea Control Concept. In November 1973, the squadron moved from Lakehurst, New Jersey, to Jacksonville, Florida and soon joined Helicopter Antisubmarine Wing ONE, thus closing its Sea Control chapter.

As a Carrier Air Wing squadron, HS-15 has operated aboard seven East Coast carriers: USS NIMITZ (CVN 68), USS AMERICA (CV 66), USS INDEPENDENCE (CV 62), USS FORRESTAL (CV 59), USS SARATOGA (CV 60), USS ENTERPRISE (CVN 65), USS EISENHOWER (CVN 69), and is presently aboard USS GEORGE WASHINGTON (CVN 73), as an element of Carrier Air Wing SEVENTEEN (CVW-17). Over the years, operations have taken HS-15 to the Atlantic, Arctic, and Indian Oceans as well as the Adriatic, Arabian, Caribbean, Mediterranean, and Norwegian Seas. During these many deployments the Red Lions have played a unique role in U.S. foreign policy and military actions. Involvements have included: Iranian/ Afghanistan Contingency Operations,

the U.S. Multinational Peacekeeping Force in Lebanon, combat operations during Operation URGENT FURY in Grenada, extensive North Atlantic ASW operations with NATO, Operation PROVIDE COMFORT in Iraq, and Operation PROVIDE PROMISE/DENY FLIGHT in the former Yugoslavia.

While conducting peacekeeping operations off the coast of Lebanon in 1982, the Red Lions were specifically tasked with shuttling U.S. Special Envoy Phillip Habib to and from Beirut as he negotiated a settlement to the Israeli-PLO conflict.

During USS INDEPENDENCE (CV 62) 1983-84 "Combat Cruise," HS-15 was deployed as an element of CVW-6. Throughout the hostilities in Grenada, the Red Lions provided combat SAR service as well as small boat interdiction patrols while operating a forward detachment from the deck of the USS MOOSBRUGGER. During this detachment, HS-15 rescued eleven wounded personnel from a downed Army helicopter. After a short turnaround cycle, the squadron returned to the Mediterranean Sea and Indian Ocean for a last deployment on USS INDEPENDENCE before reassignment to USS FORRESTAL with CVW-6. Embarked in USS FORRESTAL, the squadron deployed to the Mediterranean Sea in June 1986, and in August 1987 to the Norwegian Sea as participants in Ocean Safari '87.

The squadron's 1988 deployment took the Red Lions to the Mediterranean Sea through the Suez Canal, to the North Arabian Sea and the Gulf of Oman. This presence was to ensure the unobstructed passage of ocean commerce through the Strait of Hormuz into the Persian Gulf. This phase of the deployment included an impressive 108 consecutive days at sea. Among the high points of the deployment was the opportunity to meet the Commander-in-Chief, President George Bush, prior to his "Summit on the Sea" with Soviet President Mikhail Gorbachev.

With the commencement of Operation DESERT SHIELD in August 1990, and then DESERT STORM in January 1991, CVW-6 and HS-15 entered a grueling training regime in preparation for possible deployment to the Middle East. In 24 days, vice the normal 4 months, CVW-6 completed Refresher Training and Advance Phase, finishing the most intensive carrier work-up period ever conducted in the Atlantic Fleet. USS FORRESTAL, CVW-6 and HS-15 deployed to the

Mediterranean Sea in May 1991 in support of Operation PROVIDE COMFORT.

The Red Lions returned to NAS Jacksonville in late December, just 4 days before Christmas. Shortly thereafter, USS FORRESTAL was decommissioned from active duty and HS-15 left CVW-6.

In the spring of 1992, HS-15 began the transition to the SH-60F Seahawk helicopter. After traveling to NAS North Island in San Diego, California to undergo training for the new aircraft, the squadron returned in late 1992 and completed transition in March 1993.

As new members of the CVW-17/ USS SARATOGA team, the Red Lions returned to sea in June 1993 in preparation for a January 1994 deployment, which would be the last for USS SARATOGA (CV 60). The Red Lions were quick to demonstrate their new and vastly expanded capabilities with the SH-60F and HH-60H helicopters.

One hundred percent of the aircrews were Night Vision Goggle qualified, thereby enhancing their night ASW and SAR capabilities. Their integrated CSAR training with SEAL Team EIGHT formed the most operationally ready Combat Search and Rescue Team in the U.S. Navy. HS-15 and SEAL Team EIGHT stood over 1000 hours of CSAR alert in the Adriatic Sea supporting Operations PROVIDE PROMISE and DENY FLIGHT. Additionally, HS-15 conducted intensive Vertical Replenishment operations for the CV 60 Battle Group.

In fall of 1994, USS SARATOGA was decommissioned. CVW-17 and HS-15 moved on to the USS ENTERPRISE (CVN 65) in September 1994. The Red Lions completed a Mediterranean Sea/ Persian Gulf deployment in December 1996. Over the course of that cruise, the squadron flew almost 2000 hours and was involved in Operations DECISIVE ENDEAVOR and SOUTHERN WATCH, and participated in JUNIPER HAWK with the Israeli military.

The year 1997 was a watershed year for the Red Lions, as the squadron became a part of the USS DWIGHT D. EISENHOWER (CVN 69) battle group and welcomed its first female pilots to the wardroom. HS-15 honed its skills in anti-submarine warfare during detachments to AUTEC, in the Bahamas, and contributed to the evolution of Special Warfare helicopter operations by training several pilots in the Mountain Flying curriculum offered at NAS Fallon, Nevada.

In February of 1998, the Red Lions were once again on board the EISENHOWER for their most demanding exercise yet, COMPTUEX. This evolution included integrated training for CVW-17, culminating with a three-day war near the end of the period. COMPTUEX would also prove to be a test of HS-15's SAR capabilities, as the Red Lions carried out two successful rescues at sea. The first one involved an F/A-18 pilot, who ejected after a faulty catapult shot, while the second one included the nighttime transfer of a sick individual to a shore hospital.

By late April 1998 the Red Lions were ready for their six month deployment, and boarded the IKE to make the trek to the Persian Gulf with the rest of Carrier Air Wing 17. Plans changed and the carrier battle group received new orders to deploy in June to enforce Operation DELIBERATE GUARD, patrolling the Adriatic Sea off the coast of the former Yugoslavia. During the deployment, the Red Lions logged over 2500 hours of flight time and participated in multiple detachments. Their cruise had them staying in the Mediterranean for the bulk of the six month deployment, visiting such ports as Corfu, Antalya (Turkey), Rhodes, Naples, and Cartagena, Spain.

At mid-cruise, the Red Lions were at the peak of combat readiness for the crisis in Kosovo. With rising political tensions between Serbia and NATO, combined with the threat of the Sava class submarine, HS-15 stood ready with both a Combat Search and Rescue package and a fully armed USW response. Once released from the Adriatic, the Red Lions headed to the Arabian Gulf to provide CSAR support to the air wing and the U.S. Air Force during rising tensions with Iraq.

HS-15 was also there when lives depended on them, conducting five open ocean rescues of downed Naval Aviators, two medical evacuations from U.S. submarines at sea, and 13 medical evacuations from U.S. and foreign ships at sea.

The year 1998 ended with HS-15 receiving the HS Wing Maintenance Trophy for excellence, the Armed Forces Service Medal, and the Armed Forces Expeditionary Medal. The Red Lions also received the coveted Battle "E" for battle efficiency, the Safety "S" for an impeccable record in the safety department, and was also the runner up for the Thatch Award for ASW excellence.

In 1999, the Red Lions joined the USS GEORGE WASHINGTON (CVN 73) Battle Group and began the process of preparing for sea again. HS-15 provided Search and Rescue capabilities for nine fixed wing carrier qualification detachments. The Red Lions also journeyed to AUTEC three times, enhancing USW training and earning torpedo qualifications, allowing the squadron to win the HS Wing ASW pro week.

In September, the Red Lions won the coveted Arleigh Burke Award for the most improved squadron, ship, or submarine in the entire U.S. Atlantic Fleet, for their outstanding performance in 1998.

In December 1999, the Red Lions began workups and played a critical role in the successful completion of TSTA I and II aboard the USS GEORGE WASHINGTON. These exercises evaluated the ship's readiness for the upcoming June 2000 deployment, and provided HS-15 with ample opportunity to train in a shipboard environment.

After the winter holidays, HS-15 was on its way once again, this time to NAS Fallon, Nevada. In this isolated locale, HS-15 focused on the Combat Search and Rescue (CSAR) aspect of its mission. The Red Lions completed a large number of flights geared towards familiarizing its pilots and aircrew with desert and mountain terrain flying. The Red Lions also worked extensively with the SEALs, including inserts and extracts, gunnery exercises, and Non-combatant Evacuation Operations (NEO).

HS-15 is currently preparing for deployment aboard the USS GEORGE WASHINGTON in June 2000.

PATROL AND RECONNAISSANCE WING ELEVEN

The history of Patrol and Reconnaissance Wing ELEVEN has been long and colorful. Commissioned as Fleet Air Wing ELEVEN (FAW-11) on August 15, 1942, at Norfolk, Virginia, the Wing moved five days later to San Juan, Puerto Rico. It was charged with providing anti-submarine protection for shipping convoys in the southwest Atlantic and the Caribbean at the height of the German U-boat campaign.

The PBY-5A aircraft, with only limited capability for spotting and destroying enemy submarines, patrolled a million square miles of ocean with amazing success. The three assigned squadrons were credited with sinking ten German submarines and damaging eighteen others. Only two aircraft were lost.

Peacetime brought a winding down of the Caribbean operations, and with the closing of facilities at San Juan in 1950, FAW-11 shifted its homeport to NAS Jacksonville, arriving November 24, 1949. From this base, flying the multi-engine P2V Neptune aircraft, Wing squadrons continued to patrol vast ocean areas, participate in Fleet exercises, and train for future commitments.

These commitments, however, took on new dimensions as the capabilities of patrol aircraft expanded. Anti-submarine warfare (ASW) flights were still the first order of the day, but secondary missions of aerial mine warfare, search and rescue, and photographic intelligence were assigned.

In 1964, Wing squadrons began transitioning to the new P-3A Orion aircraft. By December 1969, all Wing ELEVEN squadrons were operating the Orion, and the 25-year career of the venerable P2V came to an end. In July 1971, the art of ASW reached a new level of sophistication when the squadrons began transitioning to the new computerized P-3C version of the Orion.

The name was changed from Commander, Fleet Air Wing ELEVEN to Commander, Patrol Wing ELEVEN on June 30, 1973. In 1998, Wing ELEVEN assumed the duties as administrative ISIC for Fleet Air Reconnaissance Squadron TWO (VQ-2) based in Rota, Spain. Answering the need for increased Intelligence, Surveillance, and Reconnaissance (ISR) assets, the MPRA community also implemented the Anti-Surface Warfare Improvement Program (AIP). These two actions brought about the command's recent name change to include the reconnaissance aspect. COMPATRECONWING ELEVEN now consists of four squadrons (VP-5, VP-16, VP-45, VQ-2), 42 aircraft, and approximately 1650 men and women. Wing squadrons operate the highly complex P-3C Update III Orion and EP-3E Aries II aircraft.

Patrol and Reconnaissance Wing ELEVEN squadrons have served with distinction in the Korean conflict, Lebanon, Berlin, and Cuban missile crises, Operations DESERT SHIELD/STORM in the Persian Gulf, and more recently, Operations ALLIED FORCE and DECISIVE ENDEAVOR in the Balkans, and Operations NORTHERN/SOUTHERN WATCH over Iraq. Squadrons deploy to Roosevelt Roads, Puerto Rico; Keflavik, Iceland; Sigonella, Sicily; Incirlik, Turkey; and Souda Bay, Crete, Greece. They also help train foreign Navies in P-3 operations and routinely take part in joint operations with British, French, Dutch, Norwegian, Canadian, and other multi-national forces.

LOCKHEED P-3 ORION AND EP-3E ARIES II

The P-3C Orion flown at NAS Jacksonville by Patrol (VP) squadrons was developed from the Electra civilian airliner and is powered by four turboprop engines. The standard tactical crew of 11 consists of three Pilots, a Tactical Coordinator, a Navigator/Communicator, two Flight Engineers, two Acoustic Operators, one Non-Acoustic Operator, and an Inflight Technician.

The P-3C is recognized throughout the world as the premier anti-submarine warfare (ASW) platform; it is unequaled in its ability to locate, track, and, if required, attack hostile submarines beneath the waves. The Maritime Patrol community has successfully demonstrated its ASW capabilities in every ocean of the world.

Once primarily an ASW platform, the P-3C's mission has expanded to include anti-surface strike, drug interdiction, surveillance, and strategic blockades to enforce United Nations imposed sanctions. Other important missions include intelligence and data collection, humanitarian assistance, and search and rescue.

The Orion is capable of delivering a host of weapons, to include MK-46 and MK-50 torpedoes, Harpoon, Maverick, and SLAM missiles, Rockeye, and bombs. The aircraft's multi-mission flexibility also includes a pinpoint mine laying capability in harbors and shipping lanes.

The success of these operations is made possible by the array of sophisticated communications, navigation, detection, and monitoring systems installed on the P-3C platform. A sensitive acoustic system monitors active and passive sonobuoys. Non-acoustic detection systems include the APS-115 radar, Infrared Detection System (IRDS), Magnetic Anomaly Detector (MAD), and ALR-66 ESM equipment. Crews employ gyro-stabilized binoculars, night vision goggles, stabilized video systems, and a remarkably sophisticated long-range electro-optical system to supply battle group commanders with useful real-time information.

In addition, new system upgrades add such equipment as the Inverse Synthetic Aperture Radar (ISAR) imaging system to already potent P-3C capabilities. The Anti-Surface Warfare Improvement Program (AIP) delivers computer and communication enhancements, improved electro-optical systems, Synthetic Aperture Radar (SAR), and survivability systems.

The EP-3E ARIES II aircraft is the electronic warfare and reconnaissance variant of the P-3 aircraft and is flown by VQ-2 in Rota, Spain. It uses state-of-the-art electronic surveillance equipment for its primary mission. There are 24 numbered seating positions, of which 19 are crew stations. The ARIES II is capable of a 12+-hour endurance and a 3000+ nautical mile range. The normal crew complement is 24, 7 officers and 17 enlisted aircrew. The EP-3E typically carries three pilots, one navigator, three tactical evaluators, and one flight engineer. The remainder of the crew is composed of equipment operators, technicians, and mechanics.

The multi-mission P-3 possesses limitless potential with rapid response to short-notice tasking, and the ability

to conduct large open ocean searches with a long endurance capability. Interoperable with numerous platforms and task forces, the P-3 is an invaluable asset in a multitude of employment scenarios.

SQUADRONS

PATROL SQUADRON FIVE

Originally known as the "Blind Foxes", Patrol Squadron five (VP-5) was commissioned at Whidbey Island, Washington, in September 1942 as BOMBING SQUADRON 135. It was nicknamed the BLIND FOX squadron and was assigned the PBY "Catalina" aircraft. In less than a month, the squadron received a new aircraft, Lockheed's PV-1 "Vega Ventura," which was flown in several bombing missions during World War II. In 1948, the squadron received its first Lockheed P2V "Neptune" aircraft which contained the Magnetic Anomaly Detector (MAD). Shortly thereafter, the squadron became known as the Mad Foxes and in December 1948, was designated Patrol Squadron Five.

Jacksonville, Florida, became the permanent home of the MAD FOXES in December 1949. VP-5 aided in the recovery of America's first astronaut, Commander Alan B. Shephard, Jr. on May 5, 1961, and was one of the first units ordered into action during the Cuba quarantine.

In June 1966, VP-5 transitioned to the Lockheed P3A "Orion" aircraft and attained an unequaled recognition for ASW proficiency. Specifically, VP-5 earned consecutive Battle Efficiency "E" awards in 1975 and 1976; Meritorious Unit Commendations in 1976 and 1982;

the Navy Expeditionary Medal for their support of the 1982 Lebanon evacuation; three "Hook Em" awards in 1977, 1981, and 1982; and the Arnold Jay Isbell Trophy for ASW excellence in 1982. VP-5 also received the COMPATWING ELEVEN Bronze Anchor Award and the COMNAVAIRLANT Silver Anchor Award for retention excellence in 1983.

In 1986, it was VP-5 on top of the disabled soviet Yankee class submarine in the final hours prior to its sinking. Deployed to the Bermuda sector, the MAD FOXES provided sole coverage of the event, keeping U.S. officials informed during this crucial period. The deployment earned a Silver Anchor Award for retention excellence, the Top Gunner Award for outstanding weapons proficiency, and the CINCLANTFLT Athletic Excellence Award.

While completing preparations for the 1989-90 Bermuda deployment, the MAD FOXES transitioned from the baseline P-3C to the state-of-the-art P-3C Update III.

In 1991, the MAD FOXES deployed to Rota, Spain, with extended detachments to Lajes, Sigonella, and Souda Bay in direct support of Operations DESERT SHIELD and DESERT STORM.

Departing for Keflavik, Iceland, in September 1992, VP-5 participated in operations stressing NATO interoperability and the P-3's role in coastal warfare. Upon return to Florida, the squadron was awarded the Battle Efficiency "E" for 1992.

In February 1994, VP-5 began a rigorous tri-site deployment which included "Gatekeeper" duties in Keflavik, air support for Operations Sharp Guard in the Adriatic based out of Sigonella, and a detachment in Jacksonville, Florida. The deployment accounted for 4,000 flight hours with operations spanning much of the European theater. It marked a first for U.S. operations with many Baltic states such as Poland and Lithuania.

In August 1995, VP-5 began yet another demanding tri-site deployment that split the squadron during among Keflavik, Puerto Rico and Panama. For the first time ever, one squadron met the operational requirements of the entire Atlantic Theater of Operations, previously fulfilled by two squadrons. From ASW and NATO interoperability flights up north to the drug interdiction flights down south, VP-5 completed the mission amassing over 6,000 flight hours.

In February 1997, they again deployed to Keflavik, Puerto Rico and Panama, where they conducted ASW, anti-drug ops, and training exercises. The coordinated efforts of VP-5 and other forces in the equatorial region interdicted $33 billion worth of drugs (street value) during the sixth month period. Upon arriving home in August 1997, the Mad Foxes began a demanding home cycle in preparation to being the first East Coast VP squadron to receive P-3 aircraft with the Anti-surface Warfare Improvement Program (AIP) modifications. The Mad Foxes deployed to NAS Sigonella, Sicily, in August 1998, flying missions over Bosnia-Herzegovina in support of operations Deliberate Forge - ongoing peacekeeping efforts in the region.

Safe, effective and unsurpassed in combat readiness, VP-5 has flown over 21 years accident free, including more that 100,000 flight hours.

PATROL SQUADRON SIXTEEN

Patrol Squadron SIXTEEN was established in May 1946 at Cecil Field as VP-ML-56, initially flying the amphibious PBY "Catalina." It was re-designated Patrol Squadron 741 in 1949 and operated in reserve status until hostilities began in Korea in 1951. In February 1953 the "War Eagles" were made a part of the regular Navy and designated Patrol Squadron SIXTEEN.

In the early 1960's, the squadron transitioned to the sophisticated and versatile P-3 Orion. Today the squadron is comprised of eight P3C Update III aircraft and 11 combat aircrews, each with a crew compliment of three pilots, two flight engineers, a tactical officer, a navigator/communicator, two acoustic and one non-acoustic operator, an ordnance man and an in-flight technician. Squadron manning includes an exceptionally talented group of 67 officers and 258 enlisted personnel, 25 percent of whom are women. The War Eagles were the first Navy combat squadron to integrate female aviators into combat crews.

During their history, the squadron participated in activities and exercises on both the east and west coasts throughout the world, including Space Capsule Recovery Operations of the Friendship Seven Mercury capsule of John Glenn. In 1996, the squadron was called upon in central Africa to support Operation Guardian Assistance, the humanitarian relief effort to supply aid and comfort to refugees of the war-torn region.

VP-16 has been a key contributor to successes registered in the national effort to stem the flow of illegal narcotics across our borders. During their most recent deployment to NAS Sigonella, the "War Eagles" provided real-time optical data to multinational commanders in their quest to end the violence in the Former Republic of Yugoslavia, while at the same time conducting a highly successful prosecution of a Russian submarine.

PATROL SQUADRON FORTY-FIVE

The Patrol Squadron FORTY-FIVE (VP-45) *Pelicans* of NAS Jacksonville, Florida, were initially commissioned as Patrol Squadron 205 on November 1, 1942, at NAS Norfolk, Virginia. Through the 1940's the squadron was very active, flying PBM "Mariners" in both the Atlantic and Pacific theaters. Spanning the entire globe, the *Pelicans* changed homeports 12 times in just over three years to such places as Cuba, Hawaii, Okinawa and Japan. Finally, while homeported in Bermuda in September of 1948, they were designated as VP-45. In the 1950's, VP-45 transitioned to the Martin P-5M Marlin while based at Naval Station Coco Solo, Panama.

Homeported in Bermuda from 1956-1963, the squadron assumed duties as the Bermuda Recovery Unit for the Mercury Space Program. VP-45 deployed to Cuba in 1962 for operations during the Cuban Missile Crisis and, as a result, received the Battle "E" and CNO Safety Award for outstanding performance.

The *Pelicans* began transitioning to the P-3A Orion aircraft in September 1963. The squadron became part of Fleet Air Wing ELEVEN in 1964, bringing with them the first P-3 aircraft to NAS Jacksonville. Earning a second Battle "E" and CNO Safety Award, the *Pelicans* completed the transition to the P-3A in less than eight months. Deployed to Adak, Alaska, in 1965, the squadron made history as the first Atlantic Fleet VP squadron to see action in Southeast Asia.

In May of 1968, VP-45 responded in less than two hours to the tragic loss of the nuclear submarine USS SCORPION, flying extensive search and rescue missions from Bermuda and Lajes, Azores. Later in the year, the squadron departed Jacksonville for another six month deployment in support of combat operations in Southeast Asia, operating from bases at Naval Station Sangley Point, Philippines and U-Tapao, Thailand.

The squadron began its transition to the upgraded P-3C Orion in 1972. During their five month deployment to Sigonella in 1973, the *Pelicans* were the first squadron to fly the P-3C in the Mediterranean and were awarded the Captain Arnold J. Isbell Trophy for excellence in anti-submarine warfare (ASW). During deployments to Keflavik, Iceland, in 1974 and Sigonella in 1975, the *Pelicans* conducted ASW operations and surface surveillance while participating in several NATO and SIXTH Fleet exercises. As a result, the squadron was awarded the first of five SIXTH Fleet "Hook 'em" Awards for ASW excellence.

During a Bermuda deployment in 1980, VP-45 won the coveted "Golden Wrench" award for superlative maintenance, the "Silver Anchor" award for its retention program, and the aircrews were awarded the first of five "Top Gunner" awards for accurately placing torpedoes on target. Returning to Sigonella in 1983, VP-45 crews set ASW records by logging more submerged submarine contact hours than any other squadron.

In September of 1984, VP-45 began a split deployment to Rota, Spain, and Lajes, Azores. During the five months that followed, VP-45 was awarded their fourth "Hook 'em" award for submarine contact time and third Battle "E" for overall operational excellence. The *Pelicans* deployed to Sigonella in July 1987, flying over 4,500 hours of high tempo operations in direct support of SIXTH Fleet. The *Pelicans* then returned home to Jacksonville where they became the first active duty patrol squadron to retrofit the P-3C baseline aircraft with the advanced Update III package.

The squadron deployed to NAS Bermuda in 1989, with numerous detachments between Keflavik and Roosevelt Roads. The *Pelicans* achieved unparalleled success introducing the Update III system to the Atlantic Fleet and consequently were awarded the Meritorious Unit Commendation by the Secretary of the Navy for superb achievement.

The *Pelicans* conducted another Rota/Lajes split deployment in June 1990, introducing the Update III platform to the Mediterranean. This highly successful deployment included direct support for operations SHARP EDGE and DESERT SHIELD, earning another SIXTH Fleet "Hook 'em" award and the CNO Safety Award for 1991.

In 1992, VP-45 embarked upon a split deployment between Keflavik and Jacksonville, aggressively participating in carrier battle group operations and coordinated shallow water ASW with several NATO countries. During the deployment, the squadron surpassed 155,000 hours of mishap-free flying and achieved an unprecedented 99 percent sortie completion rate. Another multi-site deployment followed in 1993, with the aircrews shoeing their versatility by participating in both operations DESERT STORM and SHARP GUARD. As a result, the *Pelicans* received the Golden Wrench award for outstanding maintenance and the Captain Arnold J. Isbell trophy for ASW excellence.

From December 1994 to June 1995, the squadron completed their first Caribbean deployment in support of the national counter drug effort and set theater records for total flight hours flown and illicit drugs seized. Working with the U.S. Coast Guard, Air Force, and Allied forces, VP-45 successfully operated from Puerto Rico, Honduras and Panama, and was awarded a Meritorious Unit Commendation for its superior performance.

The *Pelicans* deployed to Sigonella in 1996 and again set new standards for Maritime Patrol Aviation by participating in 18 exercises and detaching to five locations throughout Europe and the Middle East. The squadron again demonstrated the multi-mission capability of the P-3C, flying both tactical reconnaissance missions over Bosnia and blockade support missions in operation SHARP GUARD.

Returning to Sigonella in 1997, the *Pelicans* flew over 5000 hours with a 98 percent sortie completion rate. The sorties flown supported 18 detachments from 10 different locations, including three weeks of Search and Rescue (SAR) contingency operations out of Namibia. The *Pelicans* were the first VP squadron to bring the Standoff Land Attack Missile (SLAM) to the Med, and fired the only Maverick missile there since 1994. The squadron continued to set high standards for on-station presence and performance, both overland in operation DELIBERATE GUARD and in a remarkable 28 exercises in support of the SIXTH Fleet. For outstanding performance throughout the year, the VP-45 maintenance department received the Golden Wrench award for 1998.

The *Pelicans* departed in February 1999 on another multi-site deployment to Puerto Rico, Panama and Iceland. The squadron made history by helping seize over 3 billion dollars worth of illicit drugs in the Caribbean, continuing their ASW superiority in the North Atlantic, and intercepting the first Russian BEAR "F" aircraft in Icelandic airspace in over eight years. In July, VP-45 surpassed 30 years and 198,000 hours of mishap-free flying.

Returning home to Jacksonville in August, the squadron began another rigorous inter-deployment training cycle which includes transitioning to the newest P-3C upgrade, the Anti-Surface Warfare Improvement Program (AIP). The *Pelicans* will take AIP to the Mediterranean in August 2000.

PATROL SQUADRON THIRTY

Patrol Squadron 30 (VP-30), the "Pro's Nest," is the U.S. Navy's Maritime Patrol Fleet Replacement Squadron (FRS). VP-30's mission is to provide aircraft-specific training for pilots, naval flight officers, and enlisted aircrew prior to their reporting to the fleet. More than 750 staff personnel directly or indirectly train over 700 officer and enlisted personnel annually, utilizing 30 P-3 aircraft of various models. Foreign military personnel from Thailand, Germany, Netherlands,

Norway, Japan, and the Republic of Korea have all received specific aircrew and maintenance training on P-3 operations and systems.

VP-30 was commissioned in June 1960 at NAS Jacksonville, Florida, to train flight crews for P-5 Marlin and P-2 Neptune aircraft. In June 1963, VP-30 Detachment ALFA was formed at Patuxent River, Maryland, to begin training in the newly introduced P-3 Orion. Growth of VP-30 Detachment ALFA soon became significant enough that the squadron homeport was changed to NAS Patuxent River, Maryland, in 1966. Flight operations continued at NAS Jacksonville until P-2 aircraft were phased out of service in December 1968.

June 1969 marked the beginning of P-3C training with its computerized data processing equipment. In 1970, VP-30 assumed training for P-3 maintenance personnel with the Fleet Readiness Aviation Maintenance Program (FRAMP). From March to August 1975, VP-30 returned to its present homeport of NAS Jacksonville. In August 1991 the command was designated a Major Shore Command as the Maritime Patrol Community Fleet Replacement Squadron (FRS).

Since its establishment in 1960, VP-30 has epitomized professionalism in Naval Aviation. This trademark of the command is largely due to a rigorous process used for screening our ground and flight instructors, who come to the "Pro's Nest" with vast Fleet operational experience and undergo an extensive Instructor Under Training (IUT) syllabus prior to being assigned as instructors. VP-30 ensures the Fleet receives safe and competent replacement pilots, naval flight officers and aircrew who are ready to do the job upon reporting to their squadrons.

In 1993, VP-30 and VP-31 were consolidated into a single site FRS. There are currently 12 active duty VP fleet squadrons homeported in Brunswick, Maine; Jacksonville, Florida; Kaneohe Bay, Hawaii; and Whidbey Island, Washington, all of which are supported by the dedicated men and women of the "Pro's Nest".

VP-30's awards include three Navy Meritorious Unit Commendations for Norwegian Navy P-3B training and P-3C introduction, P-3C Update II training of the Japanese and Royal Netherlands Navies, and for the consolidation of all P-3 training into a single-site FRS; the United States Coast Guard Meritorious Unit Commendation with Operational Distinguishing Device for participation in the 1985-1986 Winter Law Enforcement Operations; the 1971, 1983, 1991, 1992, 1995 and 1998 CNO Safety Award; and the 1995 CINCLANT Golden Anchor Award for retention.

In 1998, VP-30 formed the P-3 Weapons Tactics Unit (WTU) to provide post-FRS training to the fleet. This graduate level instruction includes weaponeering, advanced tactics and weapon system employment.

VP-30 is nearing completion of Fleet introduction of the P-3C Anti-Surface Warfare Improvement Program (AIP). This program brings MPA vision 2010 into reality 11 years ahead of programmed goals. Full operational integration of this critical platform was achieved during the Kosovo conflict with AIP equipped aircraft firing the first P-3 wartime missiles since Vietnam.

NAVAL AIR RESERVE JACKSONVILLE

Naval Air Reserve Jacksonville originally formed at the station's auxiliary field at Cecil Field on June 29, 1946, has operated from Naval Air Station Jacksonville since August of 1946. The active duty headquarters staff includes over 20 officers, 200 enlisted and 24 civilian personnel.

Patrol Squadron -741 became one of the first reserve units from NARTU Jacksonville to be called to active duty during the Korean Conflict in 1951. The squadron mobilized again in 1961, bolstering the fleet for a year during the Berlin Crisis. The reserves attached to NARTU came through once again flying during the Cuban Missile Crisis.

In 1962, NARTU participated in emergency operations to aid the people of Haiti hit by Hurricane Flora, and a year later participated in rescue operations by plucking eight people from the roof of the burning Hotel Roosevelt in Jacksonville.

Many Jacksonville based units did their part during the Vietnam War, flying voluntary missions to the Western pacific in support of the U.S. efforts by airlifting humanitarian goods.

The Naval Air Reserve was reorganized into commissioned squadrons and air wings in 1970. Jacksonville became the home to Attack Squadron 203, Fleet Logistics Support Squadron 58 and Patrol Squadron 62, all based at NAS Jacksonville. Commander, Carrier Air Wing Reserve 20 was also established as the headquarters of an air wing consisting of nine tactical air squadrons located in seven states. NARTU became the Naval Air reserve Unit (NARU) at the beginning of 1972. VA-203 moved to NAS Cecil Field in 1977. The Marine Air Reserve Training Detachment, now Marine Air Group 42 Detachment, also relocated to Cecil Field in 1978, followed by CVWR-20 and its staff in 1979. Helicopter Antisubmarine Squadron 75 relocated to NAS Jacksonville from Willow Grove, Pennsylvania, in 1985. In 1996, VA-203 transitioned to the F/A-18 Hornet to become Strike Fighter Squadron 203, and became the first squadron to leave Cecil Field due to Base Realignment and Closure when they relocated to NAS Atlanta, Georgia.

Recently, VP-62 transitioned to the Charlie modification (Update III) of the P-3 Orion; HS-75 completed transition to the SH-60F helicopter in June 2000; and VR-58 is making plans to become the second Fleet Logistic Support Squadron to receive the new C-40A scheduled for Spring 2002.

Over 400 members of the NAVAIRES Jacksonville team deployed during Desert Storm.

The staff is located in Building 966 (which also houses Naval Reserve Readiness Command Region Eight Headquarters). Today, nearly 1,200 Selected Reservists drill with 12 augment units that are either directly under the command of Naval Air reserve Jacksonville or derive administrative and logistic support from the command. The command mission is to recruit, train, and retrain skilled reservists whose units augment specific fleet and shore activities for fleet operations, exercises and crisis situations. NAVAIRRES Jacksonville received the Commander, Naval Air Reserve Force 1995 Gold Helm Award for retention excellence.

FLEET LOGISTICS SUPPORT SQUADRON 58

Fleet Logistics Support Squadron 58 (VR-58) was established at NAS Jacksonville on November 1, 1977. Formal ceremonies were held in April 1978, with the delivery of the first C-9 "Skytrain" named the "City of Jacksonville". In September 1978, two additional aircraft were received, making the squadron "mission ready." VR-58 is one of seven Navy C-9B squadrons stationed throughout the United States that report directly to Commander, Fleet Joint Logistics Support Wing based at NAS Joint Reserve Base Fort Worth, Texas.

The squadron now operates four McDonnell Douglas C-9B "Skytrain II" aircraft, named after the venerable R4D transport of WWII and Berlin Airlift fame. The C-9B operates at speeds in excess of 500 mph and altitudes up to 37,000 feet and is capable of carrying six crewmembers, 90 passengers, 28,000 pounds of cargo, or various passenger/cargo loads.

VR-58 is composed of 48 officers and 224 enlisted personnel. Since its establishment, the squadron has complied 21 years and over 105,000 class Alpha mishap-free flight hours, flown more than 16 million passenger seat and cargo miles, and provided personnel and aircraft for 76 overseas detachments. Squadron missions encompass worldwide fleet support throughout the United States, Caribbean, Central and South America, Asia, Europe, Africa, the Middle East and Mediterranean, the Western pacific (including the republic of China), and Indian ocean. VR-58 was directly involved in support of U.S. forces in Lebanon and Grenada and supported NATO forces during Northern Wedding and Operation Desert Shield/Storm.

VR-58 conducts extensive training programs to ensure the highest degree of proficiency and professionalism among its pilots and aircrew, maintenance and administrative personnel. Awards include the CNO Safety Award for 1983, the Congressman Bill Chappell Award for Operational Excellence in 1993, 1994 and 1996, the Meritorious Unit Commendation in 1986, the Navy Unit Commendation in 1991, the 1995 and 1996 National Defense Transportation Association Military Unit Award, the 1995 James M. Holcombe Award for Maintenance Excellence, and the 1996 AFLSW Training Excellence Award.

The "Sunseekers" of VR-58 received the ultimate honor for the logistics community by winning the coveted Noel Davis Trophy for 1995 and 1996. Past years for the award include 1981, 1982, 1984 and 1985. The Noel Davis Award is presented to the squadron which achieves the highest level of readiness.

In February 1997, the Sunseekers stepped into naval history by flying the squadron and the Fleet Logistics community through the 100,000 hour mishap-free milestone, marking 20 years of safe squadron operations. At the completion of 1999, an additional 12,000 flight hours were added to this significant accomplishment.

VR-58 has been selected as the second squadron to receive the C40A (737-700), the replacement of the C-9B. Transition is tentatively scheduled for Spring 2002.

PATROL SQUADRON SIXTY-TWO

Patrol Squadron 62 was established at Naval Air Station Jacksonville on November 1, 1977, to provide fully manned and equipped squadrons in the event of war or national emergency. The squadron initially flew the SH-2P Neptune patrol aircraft while awaiting delivery of the P-3 Orion. Selected as the first reserve force squadron to transition to the P-3C Update III aircraft, VP-62 now operates within the same sophisticated equipment as the operational VP squadrons.

VP-62 is under the operational and administrative control of Commander, Reserve Patrol Wing. The squadron is comprised of 230 selected reserve personnel, augmented by 120 active duty personnel. Their missions, aside from the normal training load, include fleet exercise and patrol augmentation and drug interdiction flights. Naval Reserve Squadrons fly approximately 32% of the Navy's fleet maritime patrol missions.

Since establishment, VP-62 has logged thousands of operational flight hours supporting the fleet throughout the Atlantic and Mediterranean. During annual training periods, VP-62 personnel have operated out of the Azores, Bermuda, Brazil, Chile, Crete, Iceland, Norway, Panama, Peru, Portugal, Puerto Rico, Sicily, Spain and the United Kingdom. The overwhelming success of these deployments has highlighted the advanced capability of the Update III and demonstrated the Reserves ability to effectively operate and maintain front-line equipment in a challenging real world environment. In June of 1988, the Broadarrows became the first Reserve Squadron to launch a Harpoon air-to-surface missile.

Throughout the years, the "Broadarrows" of Patrol Squadron 62 have been recognized in the areas of operational readiness and command efficiency. Command awards include: Battle Efficiency "E" for 1974, 1976, 1983 and 1996; the Retention Excellence Award for 1981, 1985, 1992, 1993, 1994, 1995 and 1996; the Liberty Bell Trophy for ASW crew performance for 1978, 1983 and 1994; the Top Bloodhound Award for torpedo delivery excellence in the 1988 RESPATWINGLANT Mining Derby, 1991 and 1996; the Administrative Excellence Award for 1995, 1996 and 1997; and the AVCM Donald M.1 Golden Wrench Award for maintenance excellence in 1989 and 1994. In 2000 the Broadarrows surpassed 23 years and 77,000 hours of mishap free flying and became the first Reserve Force unit to be awarded the NAS Jacksonville Tenant Command Safety Award.

HELICOPTER ANTISUBMARINE SQUADRON SEVENTY-FIVE

Helicopter   Antisubmarine

Squadron 75 was established on July 1, 1970, at NAEC Lakehurst, New Jersey, as one of four "Citizen Patriot" HS squadrons formed during the 1970 reorganization of the Naval Air Reserve. The squadron was moved to NAS Willow Grove, Pennsylvania, in 1979 and subsequently relocated to NAS Jacksonville on October 1, 1985. HS-75 is under the operational command of Commander, Helicopter Wing Reserve located at NAS North Island, San Diego, California.

The primary mission of the "Emerald Knights" is to seek out, track and destroy enemy submarines. Constantly training for this mission, the squadron has deployed aboard several U.S. carriers, small and major combatants and auxiliaries, as well as several ship types of the Canadian Navy. The secondary mission of the squadron is to provide contributory support in the form of search and rescue and logistical support to Commander, Naval Air Reserve Force. During HS-75's 28 years of operation, the squadron has deployed aboard carriers, as well as numerous major combatants and auxiliaries. With safety always at the forefront, HS-75 has amassed over 44,000 Class "A" mishap free flight hours.

In December 1990, HS-75 responded to tasking and deployed 53 personnel and three SH-3H helicopters to NSF Diego Garcia in support of operations Desert Shield and Desert Storm. Maintaining SAR posture for 153 days, the highlight of the deployment was the rescue of three B-52 aircrew personnel on February 3, 1991. The squadron received a Meritorious Unit Commendation for its performance during Desert Shield/Storm.

In 1996, the Emerald Knights received their fourth consecutive Golden Wrench award for maintenance excellence. HS-75 also received a Meritorious Unit Commendation for its role in supporting Atlantic Fleet carrier operations during the transition to the H-60 by Commander, Helicopter Antisubmarine Wing, U.S. Atlantic Fleet. Between August 1993 and October 1998, HS-75 completed 25 detachments, embarking aboard every Atlantic Fleet carrier. In all, the squadron spent 533 days at sea and provided over 24,000 man-days in support of Training Command and Fleet Replacement Squadron carrier landing qualifications.

HS-75 has an authorized complement of 33 officers and 232 enlisted personnel who fly and maintain the unit's new SH-60F Seahawk helicopters. The SH-3 Sea Kings were phased out in December 1999. Most of the members of the squadron are Selected Reservists, who participate with the Naval Reserve as a second career and, as such, must maintain their proficiency through scheduled drills and the annual Active Duty for Training period.

SEA CONTROL WING U.S. ATLANTIC FLEET

Commander, Sea Control Wing, U.S. Atlantic Fleet is headquartered at Naval Air Station Jacksonville, Florida, and reports directly to Commander, Naval Air Force, U.S. Atlantic Fleet. It's mission is to provide U.S. Atlantic Fleet Commanders with Sea Control and Electronic Reconnaissance Squadrons fully trained and combat ready, executing all assigned tasks timely, correctly, safely and decisively.

The command was officially commissioned Air Anti-Submarine Wing ONE April 1, 1973, at Quonset Point, Rhode Island, and subsequently moved to NAS Cecil Field that fall. October 1976 marked the completion of the Atlantic Fleet transition to the S-3A Viking, replacing the propeller-driven S-2 Tracker. May 2, 1987, the command was redesignated Sea Strike Wing ONE. Following reorganization of Naval Aviation Command structures, Sea Strike Wing ONE was elevated to a major command status on October 1, 1992 and was redesignated Sea Control Wing, U.S. Atlantic Fleet. This name change reflects the new multi-mission role of the S-3B Viking and also recognizes the new command structure.

The Commander, Sea Control Wing, U.S. Atlantic Fleet is responsible for the readiness, training, administration and maintenance support of all Atlantic Fleet VS and VQ squadrons. Sea Control Wing Atlantic is comprised of five S-3B Viking squadrons: VS-22 (Checkmates), VS-24 (Scouts) VS-30 (Diamondcutters), VS-31 (Topcats), and VS-32 (Maulers) which deploy with their respective Air Wing Commanders with eight aircraft per squadron.

Sea Control Wing Atlantic is supported by an S-3 Tactical Support Center (TSC) and the local Sea Control Weapons School. The tactical Support Center is responsible for the tactical training support of VS squadrons, operational support for U. S./Foreign MPA and USAF detachment aircrew and provides tailored support to carrier ASW modules on both coasts. The Sea Control Weapons School provides simulator scheduling, instructs ground school classes and administers the Atlantic Fleet S-3B NATOPS program. The total Wing complement consists of approximately 2,000 men and women as well as 50 aircraft.

S-3 VIKING

The Lockheed S-3B "Viking" is one of the world's most versatile combat aircraft. In the course of a single mission it can seek out and destroy enemy surface and subsurface craft, provide electronic warfare support and refuel other aircraft. Its exceptional endurance, high speed dash and low speed loiter capabilities, rapid roll rate and state of the art sensors place the S-3B truly in a class of its own. Crewed by a Pilot and two Naval Flight Officers, the S-3B can deploy a wide variety of ordnance including anti-ship missiles, torpedoes, bombs, rockets, mines, and flares. Since receiving its first S-3 in 1976, VS-30 has operated the aircraft through numerous deployments and thousands of flight hours. With continued upgrades, the S-3 will continue serving well into the next century on the front lines of U.S. Naval operations around the globe.

SEA CONTROL SQUADRON 22

The squadron traces its roots to VA-22, the first east coast carrier based ASW squadron. VS-22 was established on May 18, 1960, at Naval Air Station, Quonset Point, Rhode Island, concurrently with the establishment of Carrier Antisubmarine Air Group FIFTY-FOUR. Since inception, the "Checkmates" have operated from the decks of legendary carriers that include: USS ESSEX (CVS-9), USS INTREPID (CVS-11), USS RANDOLPH (CVS-15), USS WASP (CVS-18), USS ORISKANY (CVA-34), USS LAKE CHAMPLAIN (CVS-39), USS SARATOGA (CV60), USS INDEPENDENCE (CV-62), USS KITTY HAWK (CV-63), USS ENTERPRISE (CVN-65), USS AMERICA (CV-66), USS JOHN F. KENNEDY (CV67), USS DWIGHT D. EISENHOWER (CVN-69), and USS THEODORE ROOSEVELT (CVN-71).

From 1960 until 1974, VS-22 flew the venerable Grumman S-2 "Tracker" best known as the "Stoof." The squadron now flies the sophisticated S-3B "Viking" aircraft built by Lockheed California Company.

The "Checkmates" have set the VS community standards since the squadron's inception. The highlight of 1961 was the recovery of America's first astronaut, Alan B. Shepard, after his pioneering space flight on May 5 of that year. In August 1965, VS-22 embarked in USS LAKE CHAMPLAIN for another space capsule recovery, this time Gemini 5 with astronauts Gordon Cooper and "Pete" Conrad. Returning to USS ESSEX, VS-22 participated in the recovery of Apollo 7 with astronauts Shirra, Eisele, and Cunningham.

On April 1, 1973, VS-22 joined Air Antisubmarine Wing ONE, the precursor of Sea Control Wing, U.S.

Atlantic Fleet. On November 8, 1973, with the subsequent closure of NAS Quonset Point, Rhode Island, VS-22 transferred to NAS Cecil Field, Florida.

Jet transitions started in the summer of 1974 with pilots training in T2s and A-4s at NAS Meridian, Mississippi. On January 6, 1976, after 22 years of flying the S-2 Tracker, VS-22 sailed from NAS Mayport, Florida, to the Mediterranean Sea onboard USS SARATOGA as the first deployed S3A Viking squadron.

In August of 1990, due to the Iraqi invasion of Kuwait, the "JFK" responded with a No-Notice deployment to the Red Sea. "Checkmate" aircrews flew the first ever Commander, Battle Force Red Sea Iraqi Border surveillance and signals collection flights.

Operation DESERT STORM commenced the liberation of Kuwait on January 17, 1991. "Checkmates" flew over 1100 combat hours and 324 combat sorties in direct support of coalition forces. Target information gleaned by VS-22 aircraft played a major role in the suppression of enemy air defenses during the first days of DESERT STORM. From January 22, 1991 until the cease-fire on February 28, 1991, the "Checkmates" flew in every CVW-3 strike against Iraq.

VS-22 tactical innovation did not end with the war. Support of CJTF-4 Counter Narcotics Operations during the last part of 1991 through the first part of 1992 earned the "Checkmates" the Joint Meritorious Unit Award for identifying over 1500 contacts, confirming 50 as suspected offenders.

After a long pre-deployment work-up schedule, VS-22 deployed on MED 1-93 embarked in USS JOHN F. KENNEDY in October 1992. Concentrating on multi-national Mediterranean and Adriatic Sea exercises during the first half of deployment, VS-22 provided initial in-flight refueling training for Egyptian Mirage 2000 pilots and practiced ASUW skills while leading over 40 multi-national, integrated and air wing exercise strikes against NATO ships.

Operations PROVIDE COMFORT and PROVIDE PROMISE brought a shift in VS-22's role as, once again, the S-3B Viking's electronic support systems became the Battle Group Commander's eyes and ears in another potentially hostile environment. While conducting air wing proficiency operations during PROVIDE PROMISE, the "Checkmates" provided invaluable radar locating and Command and Control information

while U.S. Air Force assets dropped relief supplies throughout a fluid electronic warfare environment. Additionally, VS-22 contributed to the U.S. Navy's evolving focus on "From the Sea..." with near land ESM, ISAR, and CCC missions.

In February of 1994, VS-22 began in earnest with Carrier Air Wing three and the crew of USS DWIGHT D. EISENHOWER. It was at this time that Congress passed laws permitting the embarkation of women in combat units. VS-22 had the distinction of being the first S-3B command to be assigned female Sailors. On April 16, 1995, VS-22 returned from their Mediterranean deployment onboard USS DWIGHT D. EISENHOWER.

On November 26, 1996, the "Checkmates" commanded by Scott Albertson, deployed with CVW-3 onboard USS THEODORE ROOSEVELT to the Mediterranean/North Arabian Gulf. VS-22 participated in Operations SOUTHERN WATCH and DECISIVE ENDEAVOR as well as numerous other joint exercises. On April 21, 1997, in the Eastern Mediterranean, the "Checkmates" became the first S-3B squadron to launch the AGM-65F Infrared Maverick missile.

Most recently, VS-22 deployed onboard USS ENTERPRISE. During JTG 99-1, the Checkmates distinguished themselves in Operation DESERT FOX. Over the northern Arabian Gulf and Kuwait, VS-22 excelled as a critical strike support asset, as Naval Air Forces dominated the Iraqi skies from the deck of ENTERPRISE. During OPERATION JUNIPER STALLION, VS-22 became the first S-3B squadron to fire a live AGM-65F Maverick missile against a land target. The crew scored a direct hit.

Operating in the Adriatic, VS-22 participated in OPERATION NOBLE ANVIL, providing EW support and much needed fuel to overland strikers.

VS-22 is currently deployed onboard USS Harry S. Truman.

The "Checkmates" have won ten Battle Efficiency Awards. The 1981 Award made VS-22 the first East coast S-3A squadron to win back to back Battle "E"s. Other squadron awards include eleven COMNAVAIRLANT Aviation Safety Awards, ten Captain Arnold J. Isbell Trophies for ASW Excellence, one Thatch Trophy, three Navy Unit Commendations, five Meritorious Unit Commendations, one Silver and four Golden Wrenches for maintenance excellence, and one Golden and three Silver Anchors for superior retention.

Sea Control Squadron TWENTY-FOUR (VS-24) traces its history back to January 1, 1943, when it was commissioned as Bombing Squadron SEVENTEEN (VB-17). The squadron embarked aboard USS BUNKER HILL (CVS-17) and USS HORNET (CVS-12) for action throughout World War II.

The post-war period was one of great changes for the squadron. Its designation and tasking changed several times until April 8, 1960, when it was designated Air AntiSubmarine Squadron TWENTY-FOUR and assigned the AntiSubmarine Warfare (ASW) mission. In 1993, the squadron was redesignated Sea Control Squadron TWENTYFOUR, which reflected the new multi-mission role of the Viking aircraft.

Throughout the years, VS-24 has flown five different types of aircraft from fifteen aircraft carriers. The last airframe change for the Scouts occurred in August 1975 when the squadron transitioned from the prop-driven S-2G Tracker to the turbofan S-3A Viking. The squadron upgraded to the S-3B in September 1989. This modification added the capability to launch the Harpoon missile, added Inverse Synthetic Aperture Radar (ISAR), provided the ability to deploy electronic countermeasures and aerial refuel other aircraft, and improved the Electronic Surveillance Measures (ESM) and acoustic detection systems.

Today this squadron exists to support the accomplishment of the Navy's missions of forward presence, power projection, sea control, and on-scene crisis response. Specifically, VS-24 provides the operational commander with advanced command and control, surveillance, and battle space dominance. The Scouts accomplish this by ensuring the highest state of readiness and providing an all weather

sea strike and sea control platform that is highly capable in all warfare areas. Sea Control Squadron TWENTY-FOUR is an integral element in the successful execution of Anti-Surface Warfare, Electronic Warfare, Over the Horizon Targeting, Counter Targeting, and Organic Refueling.

In 1991, the Scouts deployed to the Arabian Gulf in support of Operation DESERT STORM, VS-24 crews were responsible for the first S-3B land strike over enemy territory and the first sea strike against a hostile patrol craft. Following the war, Sea Control Squadron TWENTY-FOUR participated in Operation PROVIDE COMFORT rendering aid to Kurdish refugees fleeing Iraq.

In 1995, the Scouts deployed aboard USS THEODORE ROOSEVELT (CVN-71) for operations in the Red Sea and Arabian Gulf supporting Operation SOUTHERN WATCH, and in the eastern Mediterranean in support of NATO strikes against Bosnian-Serb military targets. During Operation DELIBERATE FORCE, VS-24 became the first S-3B squadron to launch Tactical Air Launched Decoys (TALD) in support of overland strikes.

In 1997 VS-24 returned to the Red Sea and Arabian Gulf to support the ongoing efforts of Operation SOUTHERN WATCH, and participated in Operation DELIBERATE GUARD in the Mediterranean and Adriatic Seas. During this deployment, the Scouts were the most heavily utilized squadron in the air wing accumulating over 2,300 flight hours while maintaining a 100 percent mission completion rate. Scout crews repeatedly proved the Viking to be the platform of choice for conducting Surface Warfare and Electronic Surveillance.

In 1999, VS-24 headed to the Mediterranean and Adriatic Seas and participated in operations in support of NATO Operation NOBILE ANVIL. The SCOUTS performed Surface Warfare and provided critical refueling for strike aircraft flying combat missions over the Yugoslav province of Kosovo. Later in the operation, VS-24's missions expanded to include Electronic Surveillance over the province of Kosovo.

Today the Scouts continue to fly missions that support all battle group warfare commanders while remaining on the cutting edge of technological advancement as the leaders of the Viking community.

During the post World War II years, TBM Avenger squadron VC-801 served as a component of the demobilized reserve Carrier ASW forces. On August 1, 1950, VC801 was re-designated VS-801 at Miami, Florida, with 18 TBM-3E Avengers. The squadron was recalled to active duty on February 1, 1951 due to the military mobilization associated with the outbreak of the Korean War and moved its home station to NAS Norfolk, Virginia.

In February of 1952, the squadron received its first AF-2S and AF-2W "Guardian" aircraft to replace the aging Avenger.

On April 1, 1953, Air Antisubmarine Squadron 801 was redesignated Air Antisubmarine Squadron THIRTY. During these early years, the squadron's emblem was developed from the original theme of the hunter "cat" stalking its prey.

VS-30 transitioned to the Grumman S-2F Tracker in October 1954. The squadron deployed on "straight deck" CVS antisubmarine carriers to the Atlantic, Caribbean and Mediterranean Sea.

In June of 1960, VS-30 changed its base of operations from Norfolk, Virginia, to NAS Key West, Florida, and was designated the S-2 TRACKER Readiness Training Squadron for the Atlantic Fleet. As the East Coast training squadron, VS-30 earned the nickname "DIAMONDCUTTERS", by taking "nuggets" fresh from the training command and honing their skills to the fine edge needed to be fleet aviators. In October 1962, VS-30 crews integrated into fleet squadrons during the Cuban Missile Crisis.

In order to consolidate all Atlantic Fleet VS assets at a single sight, VS-30 was directed to relocate its men and machines to NAS Quonset Point, Rhode Island, in July 1970. VS-30 continued its primary mission of indoctrinating and training pilots, aircrewmen, and maintenance personnel, compiling an impressive safety record of more than 50,000 accident free flight hours and earning several commendations for exemplary service.

With the closure of NAS Quonset Point, VS-30 made another permanent move in September 1973 to NAS Cecil Field, Florida. The Diamondcutters continued to train S-2 aircrews and maintenance personnel. Eventually, VS-30 took over S-2 training for both the Atlantic and Pacific Fleets as well as training for crews from Argentina, Turkey, Brazil, Venezuela, Peru and South Korea.

On April 1, 1976, after 22 years of flying the S-2 TRACKER, VS-30 became an operational fleet squadron and transitioned to the S-3A VIKING. The squadron received training in the maintenance and tactical utilization of their new aircraft at NAS North Island, California. The Diamondcutters began their first operational deployment with the S-3A in April 1978 aboard USS FORRESTAL (CV-59) as part of Carrier Air Wing SEVENTEEN (CVW-17). Returning from the Mediterranean in October 1978, the Diamondcutters began preparing for their second Mediterranean Deployment with the S-3A, which began in November 1979. During this deployment, VS-30, CVW-17, and the USS FORRESTAL were members of the first Carrier Battle Group since World War II to be the only Carrier Battle Group in the Mediterranean. The Diamondcutters returned in May 1980 and deployed again to the Mediterranean aboard the USS FORRESTAL in March 1981. This third deployment included operations in the Mediterranean Sea and the Indian Ocean with support of the Battle Group's participation in the Palestinian evacuation of Lebanon.

VS-30's fifth Viking deployment was in April 1984 aboard USS SARATOGA (CV 60) to the Mediterranean Sea. This deployment included the major NATO exercise DRAGON HAMMER where the Diamondcutters continued to reach new flight hour and sortie completion milestones.

After a ten-month turnaround, the Diamondcutters deployed again in August 1985 to the Mediterranean Sea and Indian Ocean aboard USS SARATOGA on the most eventful cruise

in its history. This deployment included operations in support of national tasking during the ACHILLE LAURO hijacking crisis, the first ever night transit of the Suez Canal by a U.S. warship, and the crossing of the equator for the first time in squadron history. During this period VS-30 was the only S-3A squadron deployed from the East Coast. Later in the deployment, the squadron provided continuous surveillance support during three freedom of navigation operations in the vicinity of Libya.

The squadron returned from the seven and one-half month deployment in April 1986 and, following a fourteen-month turnaround, again deployed to the Mediterranean Sea aboard USS SARATOGA. During this highly successful deployment VS-30 was awarded the COMSIXTHFLT "HOOK 'EM" Award for ASW excellence and was nominated for the Arleigh Burke Fleet Trophy. VS-30 returned to NAS Cecil Field in November 1987. In February 1988, VS-30 was awarded its first COMNAVAIRLANT Battle Efficiency Award and its third CNO Aviation Safety Award. In May 1988, VS-30 participated in a highly successful joint exercise with the Colombian Navy. Following the squadron's return from Colombia, VS-30 was awarded its first CAPT Arnold J. Isbell Award and the Admiral John S. Thatch Award for ASW excellence.

In July 1988, VS-30 became the first fleet squadron to receive the enhanced capability Harpoon/ISAR equipped S-3B. In August 1988, the squadron deployed aboard USS INDEPENDENCE (CV 62) for the "Around the Horn" cruise.

On 7 August 1990, the Diamondcutters deployed with Carrier Air Wing Seventeen onboard USS SARATOGA to the Red Sea as part of the UN Coalition Forces in direct support of Operations DESERT SHIELD/DESERT STORM. The S-3B was used in every facet of naval forces tasking during maritime interdiction and in subsequent strike operations against Iraq during Operation DESERT STORM. VS-30 returned to NAS Cecil Field after a seven and one-half month deployment during which squadron aircrews flew in more than 44 strikes and compiled 15230 flight hours in 258 combat sorties.

VS-30 deployed on USS SARATOGA again on May 5, 1992, the squadron's ninth Viking cruise. The ensuing six month Mediterranean

deployment included large scale NATO exercise, DISPLAY DETERMINATION. VS-30 had a banner year in 1992 - in addition to flying 4,188 flight hours, the squadron won its second COMNAVAIRLANT Battle Efficiency Award, its fourth CNO Aviation Safety Award, and its second CAPT Arnold J. Isbell Award for ASW excellence.

On 12 January 1994, the Diamondcutters embarked aboard USS SARATOGA for her farewell deployment to the Mediterranean Sea. Again VS-30 provided large-scale support to NATO forces involved in peacekeeping efforts in the Adriatic Sea off the coast of Bosnia-Herzegovina. While supplying the SARATOGA Joint Task Group with a variety of vital missions in support of Operations DENY FLIGHT, PROVIDE PROMISE, and SHARP GUARD,, the Diamondcutters distinguished themselves by collecting vital Electronic Intelligence, assuming Surface Combat Air Patrol responsibilities, and providing direct support for strike elements as airborne tankers. VS-30 also participated in numerous joint NATO and bi-lateral exercises including NOBLE STALLION, DYNAMIC IMPACT, and ILES D'OR, operating with forces from more than a dozen countries. After completing 1,620 fours of flight and over 660 traps in MED 1-94 deployment, VS-30 returned home having earned a second COMSIXTHFLT "HOOK-EM" Award for USW excellence and the CVW-17 1-94 squadron "TOP HOOK" Award.

June 28, 1996 marked the Diamondcutters first deployment aboard the USS ENTERPRISE (CVN 65). The ENTERPRISE had just finished going through an extensive five year overhaul in the Newport News shipyards. This Med deployment was marked by operations in both the Mediterranean and the Arabian Gulf. Aircrew flew in support of Operation DECISIVE ENDEAVOR with NATO forces in the former Yugoslavia. After a record setting high-speed transit through the Suez Canal and around the Arabian Peninsula,, VS-30 participated in Operation SOUTHERN WATCH over southern Iraq along with the USS CARL VINSON (CVN-70). It was the first time that more than one aircraft carrier had been in the Gulf since the Gulf War in 1991.

The Diamondcutters completed an intense work-up cycle on May 6th, 1998 which included several detachments to NS Roosevelt Roads, Puerto Rico, NAS Fallon, Eglin AFB, and seven at sea periods on the USS DWIGHT D. EISENHOWER (CVN-69). On June 10th, 1998 the squadron deployed with CVW-17 on the USS DWIGHT D. EISENHOWER. While on deployment, the squadron participated in several multi-national exercises including SHAREM 125 and FANCY '98 in the Mediterranean, Operation DELIBERATE FORGE in the Adriatic, and Operation SOUTHERN WATCH in the Arabian Gulf Perhaps more importantly, the squadron used its time on deployment to focus some of its tactical efforts on several weapon systems, which were new to the Sea Control community. The new systems included the AGM-65F Maverick missile and the AWW-13 Data Link Pod which can be used to control data link weapons such as the AGM84E Standoff Land Attack Missile (SLAM) and the AGM-62 Walleye.

The squadron achieved their primary objective of developing and implementing a training syllabus on the Maverick missile, and successfully attained a 100% qualification rate among crewmembers. Along with the training syllabus, combat checklists were created to accompany the successful tactics that were being perfected for the Maverick.

The VS-30 Diamondcutters went on to finish their deployment and recently returned to NAS Jacksonville on December 10th. The squadron was awarded its third COMSIXTHFLT "HOOK-EM" award for USW excellence in the Mediterranean theater and its third CAPT Arnold J. Isbell award and the 1998 CSCWL Arleigh Burke award.

The Diamondcutters spent the early portion of 1999 engaged in turnaround training for their June 2000 deployment aboard the USS GEORGE WASHINGTON (CVN-73). This deployment will mark the first time a Viking squadron has cruised with the new Carrier Airborne Inertial Navigation System II (CAINS II) which includes the Electronic Flight Instruments (i.e. glass cockpit display).

In December of 1999, VS-30 achieved a very noteworthy milestone. The squadron met the requirements to receive the Naval Safety Center Twenty-Year Class "A" Mishap Free Award. VS-30 also was nominated for the "PHOENIX" award for maintenance efficiency in 1999.

While onboard USS GEORGE WASHINGTON (CVN-73) for Comprehensive Training Exercise (Comptuex) in March of 2000, the Diamondcutters were awarded the Sea Control Wing Atlantic (CSCWL) Conventional Weapons Award for 1999.

In the ever evolving world of naval aviation, VS-30 has helped keep the S-3B Viking on the cutting edge. They have accomplished this by adapting to a changing enemy threat and incorporating new weapons and technology. For minimal cost and man-hours, the Viking has been a premier ASW platform and, for the last ten years, an even more effective Surface Warfare asset. The S-3B continues its transformation by taking on a new role in Strike Warfare through the addition of weapons like the Maverick and the control of others like the Walleye and the SLAM.

SEA CONTROL SQUADRON 31

Sea Control Squadron 31 (VS-31) originally existed as Scouting Squadron 31, based at Squantum Naval Air Station, Squantum, Massachusetts. The squadron served meritoriously, flying the Douglas SBD-5 Dauntless, followed by the Curtis SB2C-4E Helldiver until the end of World War II in 1945.

In April 1948, at NAS Atlantic City, New Jersey, the squadron was re-established as Composite Squadron 31 (VC-31). The following year, the squadron was redesignated Air Antisubmarine Squadron 31 (VS-31), took the name Topcats, and was subsequently moved to NAS Quonset point, Rhode Island.

In the early years, the VS-31 Topcats accomplished its antisubmarine warfare mission using TBM Avengers, followed by the AF Guardian. In 1954, the Topcats received the first Grumman S-2 Tracker. This was the first true ASW aircraft that combined both search and destroy capabilities in a single carrier-based platform.

In 1973, VS-31 departed NAS Quonset Point, its home for over a quarter of a century, and relocated to NAS Cecil Field, Florida. In the following years, the Topcats became the first VS squadron to deploy on USS Independence (CV 62) as part of the first operational test of the CV concept. This was also the last deployment the S-2 Tracker would make with the squadron.

In 1975, the jet powered S-3A Viking became the new ASW platform the Topcats would employ.

In 1988, the Topcats made their last deployment with the S-3A, and upon return from the Mediterranean, transitioned to the S-3B, a quantum improvement in antisubmarine and anti-surface warfare capabilities. In March 1990, VS-31 became the first squadron to deploy with the S-3B.

Throughout the years, the Topcats have earned a significant number of professional awards, including the Silver Anchor for retention excellence, the Admiral Jimmy Thatch Award for operational performance, and six Lockheed Corporation "Golden Wrench" awards for maintenance excellence. VS-31's excellence in antisubmarine warfare and weapons delivery continues as they have received both the 1993 "Conventional Weapons" award and the 1993 "Captain Arnold Jay Isbell" antisubmarine warfare award.

The Topcats have also received decorations in recognition of their dedication to excellence. For their efforts during the Mediterranean Deployment 2-90 and Operation "Desert Shield," the squadron received their fifth Battle Efficiency "E" and fifth Meritorious Unit Commendation. In 1990, VS-31 was awarded the Joint Meritorious Unit Award for their joint contributions to the U.S. Coast Guard's efforts to interdict illegal narcotics traffic. Med 2-94 showcased the Topcats once again as all members received two Southwest Asia Service medals for supporting United Nations operations "Southern Watch" and "Vigilant Warrior" in the Arabian Gulf. VS-31 and the rest of the George Washington Battle Group received a Navy Unit Commendation for their involvement with the ceremonies surrounding the 50th Anniversary of D-Day. Most recently, the squadron was awarded the Navy Unit Commendation and the Armed Forces Expeditionary Medal for their support of Operation "Southern Watch" in the Arabian Gulf in 1998.

The Topcats' most impressive operational achievement, however, occurred during the most recent "around the World" deployment aboard the USS John C. Stennis (CVN-74). They completed a record-breaking 1,459 consecutive sorties without missing a launch, even through four-and-a-half months of summertime heat in the Arabian Gulf. That is the highest number of consecutive sorties completed for any S-3 squadron since the aircraft entered the fleet over 25 years ago, and shows the dedication and professionalism displayed every day by each member of the VS-31 team.

The squadron relocated to NAS Jacksonville on March 31, 1998.

SEA CONTROL SQUADRON 32

Air Anti-Submarine Squadron THIRTY TWO (VS-32) was commissioned in April 1950. The squadron initially flew the TBM AVENGER under the command of LCDR Thomas B. Ellison and was based at Naval Air Station (NAS) Norfolk, Virginia. In 1951, the squadron moved to NAS Quonset Point, Rhode Island. VS-32 transitioned to the Grumman S-2F TRACKER in 1954. The VS community moved in October 1973 to their new homeport located at NAS Cecil Field, Florida. In August 1975, VS-32 began flying the S-3A VIKING. In the summer of 1990, the Maulers transitioned to the aircraft flown today, the S-3B VIKING. On 03 October 1997, VS-32 deployed for the last time from NAS Cecil Field. The Maulers returned from deployment to their new homeport of NAS Jacksonville, Florida. Since 1973, VS32 has completed three Mediterranean deployments aboard USS JOHN F. KENNEDY; shakedown operations with USS DWIGHT D.

EISENHOWER, USS NIMITZ, USS CARL VINSON, USS THEODORE ROOSEVELT and USS HARRY S. TRUMAN; and Mediterranean/Red Sea/Persian Gulf and Indian Ocean deployments aboard USS AMERICA and USS GEORGE WASHINGTON.

Since its commissioning, VS-32 has been a highly operational seagoing organization. From its first deployment in USS PALL in 1950 to its latest aboard USS JOHN F. KENNEDY (CV 67), VS-32 has been at the forefront of submarine warfare and Sea Control Warfare. Flying from the decks of USS LEYTE, USS LAKE CHAMPLAIN, USS RANDOLPH, USS JOHN F. KENNEDY, USS AMERICA, USS GEORGE WASHINGTON AND USS HARRY S. TRUMAN, the Maulers repeatedly earned their "sea legs" in the most challenging carrier operating conditions.

In recognition of its professionalism in the field of Sea Control Warfare, Commander, Naval Air Force Atlantic (COMNAVAIRLANT) awarded VS-32 the "E" for Battle Efficiency in 1960, 1964, 1969, 1982, 1986, 1993, 1995 and 1997. In 1960, 1964, 1982, 1986 and 1997 VS-32 received the Captain Arnold J. Isbell Trophy for Excellence in Anti-Submarine Warfare (ASW). In 1970, 1982, 1995 and 1997, VS-32 was honored with the Admiral Jimmy Thatch award for meritorious achievement by an ASW squadron. In 1960, 1964, 1968, 1969, 1970, 1971, 1972, 1973, 1974, 1975, 1978, 1986 and 1997, VS-32 received the COMNAVAIRLANT Aviation Safety Award. In 1986 and 1994, VS-32 received the Golden Wrench Award for maintenance excellence. VS-32 has also led the aviation community in retention statistics and was awarded the Commander in Chief, U.S. Atlantic Fleet (CINCLANTFLT) Golden Anchor Award for 1987 and 1988. In 1997, VS-32 was COMNAVAIRLANT's Silver Anchor Retention Award winner. In 1999 VS-32 was awarded the Arleigh Burke Award, given to the overall most improved squadron, and the Conventional Weapons Award, given for weapons technical proficiency.

In November 1989, upon their return from deployment, the World Famous Maulers of VS-32 transitioned to the S-3B. In September 1990 the Maulers, as part of CVW-1, participated in the first ever integrated air wing strike training exercise conducted at NAS Fallon Nevada. Emphasis was placed on the S-3B's new Electronic Surveillance Measures (ESM) system during the four-

week period. VS-32 was once again in the vanguard, this time in the early warning role.

In October 1990, the Maulers took part in pre-deployment exercises culminating in Fleet Exercise (FLEETEX) 1-91, an exercise involving mock hostilities between the Carrier Battle Group and orange forces. During this at sea period, VS-32 achieved another S-3B milestone, becoming the first carrier based S-3B squadron to launch a Harpoon missile at sea.

In February 1991, VS-32 and CVW-1 were called upon to deploy in the Persian Gulf and Arabian Sea. While in the Persian Gulf the World Famous Maulers flew Electronic Intelligence (ELINT) missions over Northern Kuwait and Saudi Arabia supporting Desert Storm operations. During Desert Storm, the Maulers became the first S-3 squadron in history to achieve an Anti-Surface kill on an enemy vessel. LCDR Bruce Bole, flying a Mauler armed surface reconnaissance flight, earned this honor and historic combat milestone.

In 1993, the VS-32 Maulers became the first S-3B squadron to participate in Adriatic over land missions, supporting DENY FLIGHT and SHARP GUARD in Bosnia-Herzegovina, UNISOM-II in Somalia, and OPERATION SUPPORT DEMOCRACY over Haiti.

The Maulers deployed on 28 August 1995 in USS AMERICA (CV 66) for the Joint Task Group (JTG) 95-3 deployment to the Mediterranean, Adriatic Sea and the Indian ocean. While on deployment, the Maulers flew in United Nations missions DENY FLIGHT, SHARP GUARD, and DELIBERATE FORCE. The Maulers participated in exercises JUNIPER HAWK with the Israelis; INFINITE COURAGE with the Italians and the British; and BRIGHT STAR with the Egyptians, French, British, and UAE.

1996 saw VS-32 participating in DECISIVE EDGE and DECISIVE ENHANCEMENT operations over the Former Republic of Yugoslavia. The Maulers also demonstrated U.S. carrier capabilities to Russian dignitaries from the Admiral Kuznetsov Battle Group showing versatility of the S-3B while extending a diplomatic and professional gesture of friendship.

With the opening of 1997, preparations were concluded for VS-32's BRAC 93 directed homeport change to NAS Jacksonville. Following the decommissioning of USS AMERICA (CV-66), Carrier Air Wing ONE was assigned to the Atlantic Fleet Carrier

USS GEORGE WASHINGTON (CVN-73). In July, the Maulers were only the second squadron to deploy with the AGM-65F Maverick Missile system. The Maulers honed this weapons system by producing safe and effective tactical employment profiles and envelopes. These were tested during War-at-Sea exercises during BRIGHT STAR with the Egyptians, French, and British. To support the new two-carrier presence requirement, the USS GEORGE WASHINGTON and CVW-17 rapidly deployed to the North Arabian Gulf. In the ensuing months, VS-32 became an integral part of Operation Southern Watch.

1998 began with the Maulers continuing their exceptional prowess in the Northern Arabian Gulf. The first three months saw the Maulers engaged in every mission of CVW-1. VS-32 ended cruise on April 3, 1998. They also saw four Orange Air support detachments both deployed and off the beach. In November, VS-32 deployed with CVW-1 onboard USS HARRY S. TRUMAN (CVN 75) for her Shakedown Cruise. While maintaining flight proficiency through unit level training and Orange Air missions, the Maulers visited the St. Croix torpedo range, qualifying all crews on their first TORPEX run. VS-32 also led CVW-1 aircraft on their largest mining event. The Maulers executed this task without error and scored top marks. Returning home in December, VS-32 enjoyed a well-earned Holiday season. The World Famous Maulers' effectiveness and efficiency has earned

them recognition throughout their community and Airwing.

1999 saw the Maulers exposed to a plethora of training exercises to prepare them for cruise in the fall. They began the year concentrating on their tactical proficiency at NAS Fallon. The Maulers then trained with CVW-1 and the USS JOHN F. KENNEDY (CV-67) in late spring during TSTA I/II. The summer was spent taking part in COMPTUEX for their blue water certification. VS-32 rounded off their last couple of months on the beach integrating with other branches of the military in JTFEX. VS-32 and CVW1 embarked in USS JOHN F. KENNEDY (CV 67) on September 22, 1999. Upon entering the Mediterranean, VS-32 took part in OPERATION BRIGHT STAR, a two day bombing exercise with the Egyptian military. Maulers proved themselves again by successfully firing their second AGM-65F Maverick missile 98 days from their initial launch. This was the first time ever an operational S-3B squadron fired Maverick off two different aircraft with the Maverick modification. The Greek Island of Avgo Nisi saw the Maulers drop MK-82 GP bombs and BDU45 500 pound inert practice bombs on time and on target. One week later the Maulers participated in another joint exercise called OPERATION BLACK SHARK. This operation was held in conjunction with the Saudi Arabians and conducted in the Red Sea. The Maulers spent the remainder of the year in support of OPERATION SOUTHERN WATCH. The total success of VS-32 for the year

of 1999 can be credited to the unprecedented teamwork of the Mauler family, both enlisted and officer alike.

SEA CONTROL WEAPONS SCHOOL

The Sea Control Weapons School supports all Viking fleet squadrons by conducting in-depth training in S-3B flight and weapons systems, their employment, and tactics, striving to ensure that each squadron deploys with its flight crews at their highest level of tactical proficiency. The Weapons School also serves as a center of excellence for the entire Sea Control community by both compiling and disseminating the latest tactics and knowledge gained from fleet experts. The Weapons School can trace its origins back to VS-27, the East Coast S-3B Fleet Replacement Squadron, which trained aircrews from 1987 until its disestablishment in September of 1994. After VS-27 was disbanded, the east coast Viking Wing stood up its own Weapons Training Unit in order to continue to provide post FRS training to Atlantic fleet squadrons. However, the dynamic evolution of Viking missions and the continual development of new weapons systems mandated the founding of dedicated and separate Weapons School, and in April of 1998, the Sea Control Weapons School was formally established at Naval Air Station Cecil Field. In June of 1999, the Weapons School was relocated to Naval Air Station Jacksonville to join the rest of the Sea Control community.

The first S-3 Viking arrives on a wet October 27, 1997.

VS-24, first S-3 squadron to arrive at NAS Jacksonville, welcoming cake. (U.S. Navy)

It has been a decade since I wrote about the future of the station. Some of the things we thought would happen did; some didn't; and some things happened we had not even envisioned 10 years ago! The helicopter community was flying the SH-3 Sea King helicopter in 1990, and plans were being made to retire all of these helicopters and replace them with the new SH-60 Sea Hawk. This came to pass for all operational squadrons based at NAS Jacksonville. In addition, the reserve helicopter squadron, HS-75, retired the last SH-3 Sea King on February 29, 2000, when, unceremoniously it was towed to the disposal yard. It will eventually end up as either a target aircraft, taken away by maybe the Air Force, or sold as so much scrap to the highest bidder. HS-75 has currently received their new SH-60 helicopters, and by the publication date of this book, will be flying them along with their operational squadron brothers. The helicopter community seems to be in a state of calm for the present, and future years of service based at NAS Jacksonville seem bright. But even now there are talks of combining this community with the on in Mayport.

The Patrol community has had a few more changes and their future platform is still not known. In 1990, it was widely thought the new Lockheed P7A would replace all of the aging P-3 Orions at the station during the 1990's. As our 1990 publication was in final review, the Navy canceled the contract with Lockheed, and I had to quickly pull all of the information concerning that plane out of our first publication. There have continued to be modifications to the P-3, and the squadrons have been reduced to half of what they were in 1990. A new hangar was built for the largest squadron in the Navy, Patrol Squadron 30, and the future of this community also looks like it will remain fairly constant for the next few years. The P-3 looks like it will also remain the aircraft of choice for this new decade.

A community that was not even considered in 1990 was the S-3 Viking squadrons. They arrived at NAS Jacksonville in 1997, and lost one squadron already as VQ-6 was disestablished. Their mission of providing antisubmarine warfare was also diminished. They now provide fueling support to the F/A-18 Hornets as their main mission. Plans concerning the new F/A-18 E and F models that are just now entering the fleet allow for extra fuel tanks and some Hornets are even being modified for fueling capabilities. Therefore, plans are for the entire S-3 Viking community to be phased out by the year 2008. The loss of this community will leave a big gap at the station, aviation wise!

Another community that may be growing that most Jacksonville and station residents know little about is U.S. Customs. Presently located in hangar 114, they now have two modified P-3 Orions with domes on top used for drug surveillance. Plans are for 10 more to come over the next two years, making a total of 12 new aircraft.

Reserve squadron VR-58 will see big changes over the next decade. Boeing just started constructing the next generation of Navy transport aircraft in late 1999. VR-58 will be the second location to replace the older C-9 Skytrains they have been flying since they were established, and replacing them with the brand new C-40A's. The first aircraft are scheduled to arrive in the Spring of 2002. One problem is their hangar roof is not tall enough for the new aircraft, so plans are being made for hangar modifications. There were even discussions on building them a completely new hangar on station.

The station itself is also changing. Although funding for new projects is scare through the next decade, some projects are on the funded list. Four new steam plants are being constructed now, and numerous runway repairs are scheduled. Additional bachelor enlisted

quarters are planned, as well as a new Administration building for the station.

Probably the biggest change will come in the business of running the station. The first fact that has been noticed by management personnel is that approximately 85 percent of the stations civilian employees are eligible to retire due to their age and/or years of service within the next five years. With the last decade of budget restrictions, the base population has grown to an average age in the late 40's, with few young personnel hired. There will be a tremendous drain of long-time knowledge within the next decade, with little young blood coming in to learn their years of knowledge to replace them! Numerous studies are also planned to have the present civilian workers "compete" for their jobs against private contractors. The first of many planned announcements for these competitions, called A-76 studies, came in early May 2000, with the second phase of announcements to occur in September 2000. Some organizations have already been through studies, and others, like the secretaries at the NADEP, are presently being studied. As funding is squeezed even more, additional cost-saving and cutting measures will happen for the next few years.

As talk of another round of base closure and realignment actions may take place in the next few years, the station will once again be fighting to keep from being closed, as will the Naval Aviation Depot.

So does all this mean NAS Jacksonville will not be here for a 75th Anniversary celebration in 15 years? No, it does not! This station, along with the Jacksonville area encompassing Naval Station Mayport and Naval Submarine Base Kings Bay, is still the most requested and preferred duty location in the Navy! The station serves a vital mission and my bet is plan now for a big celebration in 2015! There are many more years of "Service to the Fleet" in this stations future!!

COMMANDING OFFICER DATES

	Name	Dates Commanding Officer	
1.	Captain Charles P. Mason	15 October 1940 - 11 May 1942	
2.	Captain John Dale Price	11 May 1942 - 1 Mar 1943	
3.	Captain Stanley J. Michael	1 Mar 1943 - 7 Aug 1943	
4.	Captain Arthur Gavin	7 Aug 1943 - 17 Apr 1944	
5.	Captain Walton Smith	17 Apr 1944 - 27 Nov 1944	
6.	Captain Herbert E. Regan	27 Nov 1944 - 21 Jun 1945	
7.	Captain Anthony R. Brady	21 Jun 1945 - 2 Jun 1947	
8.	Commander Paul E. Emrick	2 Jun 1947 - 12 Jul 1947	(Acting*)
9.	Captain Herbert S. Duckworth	12 Jul 1947 - 15 Jun 1948	
10.	Commander F. J. Brush	15 Jun 1948 - 3 Jul 1948	(Acting*)
11.	Captain Alvin I. Malstrom	3 Jul 1948 - 15 Jun 1950	
12.	Captain D. Turner Day	15 Jun 1950 - 15 Aug 1952	
13.	Captain Burnham C. McCaffree	15 Aug 1952 - 1 Jul 1954	
14.	Captain John S. Thach	1 Jul 1954 - 31 Aug 1955	
15.	Commander John N. Myers	31 Aug 1955 - 18 Oct 1955	(Acting*)
16.	Captain William S. Harris	18 Oct 1955 - 28 Jun 1957	
17.	Commander Robert H. Smith	28 Jun 1955 - 23 Sept 1955	(Acting*)
18.	Captain Elliott W. Parish, Jr.	23 Sept 1957 - 20 Jul 1959	
19.	Captain James R. Reedy	20 Jul 1959 - 12 Aug 1961	
20.	Commander Walter J. Schub	12 Aug 1961 - 25 Sept 1961	(Acting*)
21.	Captain James R. Compton	25 Sept 1961 - 30 Jun 1964	
22.	Captain John R. Mackroth	30 Jun 1964 - 22 Jul 1966	
23.	Captain Martin J. Stack	22 Jul 1966 - 19 Jul 1968	
24.	Captain Henry O. Cutler	19 Jul 1968 - 17 Jul 1970	
25.	Captain John R. Kincaid	17 Jul 1970 - 1 Jul 1972	
26.	Captain William G. Sizemore	1 Jul 1972 - 12 Aug 1974	
27.	Captain Karl J. Bernstein	12 Aug 1974 - 14 Jul 1976	
28.	Captain Wayne D. Bodensteiner	14 Jul 1976 - 8 Aug 1978	
29.	Captain Joseph M. Purtell	8 Aug 1978 - 28 Aug 1980	
30.	Captain William C. Christenson	28 Aug 1980 - 1 Jul 1982	
31.	Captain Roger L. Rich	1 Jul 1982 - 12 Jul 1984	
32.	Captain Lynn C. Kehrli	12 Jul 1984 - 24 Jun 1986	
33.	Captain William J. Green	24 Jun 1986 - 19 Jul 1988	
34.	Captain Norman W. Ray	19 Jul 1988 - 29 Jun 1989	
35.	Captain Kevin F. Delaney	29 Jun 1989 - 26 Aug 1991	
36.	Captain Charles R. Cramer	26 Aug 1991 - 20 Aug 1993	
37.	Captain Roy D. Resavage	20 Aug 1993 - 25 Aug 1995	
38.	Captain Robert D. Whitmire	25 Aug 1995 - 8 Apr 1998	
39.	Captain Stephen A. Turcotte	8 Apr 1998 - 26 Apr 2001	
40.	Captain Mark S. Boensel	26 Apr 2001-	

No picture available since acting only.

Captain Charles P. Mason
15 October 1940 - 11 May 1942

Captain John Dale Price
11 May 1942 - 1 Mar 1943

Captain Stanley J. Michael
1 Mar 1943 - 7 Aug 1943

Captain Arthur Gavin
7 Aug 1943 - 17 Apr 1944

Captain Walton Smith
17 Apr 1944 - 27 Nov 1944

Captain Herbert E. Regan
27 Nov 1944 - 21 Jun 1945

Captain Anthony R. Brady
21 Jun 1945 - 2 Jun 1947

Captain Herbert S. Duckworth
12 Jul 1947 - 15 Jun 1948

Captain Alvin I. Malstrom
3 Jul 1948 - 15 Jun 1950

Captain D. Turner Day
15 Jun 1950 - 15 Aug 1952

Captain Burnham C. McCaffree
15 Aug 1952 - 1 Jul 1954

Captain John S. Thach
1 Jul 1954 - 31 Aug 1955

Captain William S. Harris
18 Oct 1955 - 28 Jun 1957

Captain Elliott W. Parish, Jr.
23 Sept 1957 - 20 Jul 1959

Captain James R. Reedy
20 Jul 1959 - 12 Aug 1961

Captain James R. Compton
25 Sept 1961 - 30 Jun 1964

Captain John R. Mackroth
30 Jun 1964 - 22 Jul 1966

Captain Martin J. Stack
22 Jul 1966 - 19 Jul 1968

Captain Henry O. Cutler
19 Jul 1968 - 17 Jul 1970

Captain John R. Kincaid
17 Jul 1970 - 1 Jul 1972

Captain William G. Sizemore
1 Jul 1972 - 12 Aug 1974

Captain Karl J. Bernstein
12 Aug 1974 - 14 Jul 1976

Captain Wayne D. Bodensteiner
14 Jul 1976 - 8 Aug 1978

Captain Joseph M. Purtell
8 Aug 1978 - 28 Aug 1980

Captain William C. Christenson
28 Aug 1980 - 1 Jul 1982

Captain Roger L. Rich
1 Jul 1982 - 12 Jul 1984

Captain Lynn C. Kehrli
12 Jul 1984 - 24 Jun 1986

Captain William J. Green
24 Jun 1986 - 19 Jul 1988

Captain Norman W. Ray
19 Jul 1988 - 29 Jun 1989

Captain Kevin F. Delaney
29 Jun 1989 - 26 Aug 1991

Captain Charles R. Cramer
26 Aug 1991 - 20 Aug 1993

Captain Roy D. Resavage
20 Aug 1993 - 25 Aug 1995

Captain Robert D. Whitmire
25 Aug 1995 - 8 Apr 1998

Captain Stephen A. Turcotte
8 Apr 1998 - Apr 2001

Captain Mark S. Boensel
26 Apr 2001 -

SELECTED BIBLIOGRAPHY

BOOKS

"A PICTORIAL HISTORY OF THE BLUE ANGELS," Jim McGuire, Squadrons/Signal Publications, c1981

"THE BLUE ANGELS - AN ILLUSTRATED HISTORY," CDR Rosario Rausa, Morgan Publishing Corporation

"JACKRABBITS TO JETS," Elretta Sudsbury, Hall & Ojena Publications, c1967

"HISTORY OF JACKSONVILLE, FLORIDA, AND VICINITY," T. Frederick Davis, (St. Augustine, The Florida Historical Society) 1975

"FLORIDA'S ARMY," Robert Hawk, Pineapple Press, Inc. c1986

"LOCKHEED AIRCRAFT SINCE 1913," Rene J. Francillon, Naval Institute Press, c1988

"OLD HICKORY'S TOWN - AN ILLUSTRATED HISTORY OF JACKSONVILLE," James Robertson Ward, Florida Publishing Company, 1982

"THE ILLUSTRATED ENCYCLOPEDIA OF HELICOPTERS," Giorgio Apostolo, Bonanza Books, c1984

"THE RAND MCNALLY ENCYCLOPEDIA OF MILITARY AIRCRAFT 1914-1980," Enco Angelucci, The Military Press, 1983

"U.S. NAVY AIRCRAFT 1921-1941," William T. Larkins, Orion Books, c1988

"THE WIND CHASERS - The History of the U.S. Navy's Atlantic Fleet "Hurricane Hunters"", H.J. "Walt" Walter, Taylor Publishing Company, c1992

"DICTIONARY OF AMERICAN NAVAL AVIATION SQUADRONS, Volume 1", Roy A. Grossnick, U.S. Government Printing Office, c1995

"FLORIDA: THE WAR YEARS 1938-1945" Joseph & Anne Freitus, Wind Canyon Publishing, c1998

"ADMIRAL JOHN H. TOWERS, The Struggle for Naval Air Supremacy", Clark G. Reynolds, Naval Institute Press, c1991

"JACKSONVILLE's SILENT SERVICE - A History of the Jacksonville Blood Bank and the Florida Georgia Blood Alliance, 1942-1992", Richard A. Martin, Centurian Press, c1992

"THE JACKSONVILLE STORY - 1901-1951", Carolina Rawls, Jacksonville's Fifty Years of Progress Association, c1950

"A DIFFERENT VALOR - JOSEPH E. JOHNSTON", Gilbert Govan and James Livingwood, Bobbs-Merrill Company, c1956

"THE FLIGHT JACKET, 1941" Volume VI

PAPERS:

Reports of the Adjutant General of the State of Florida, 1908, 1920-1925, 1935

The Warren Report, Appendix XIII

UNPUBLISHED PAPERS:

"THE GENESIS OF CAMP BLANDING," Brigadier General Ralph W. Cooper, Jr., (Retired), February 1978, Copy at the Jacksonville Historical Society Archives, Jacksonville University

"HISTORY OF THE U.S. NAVY IN JACKSONVILLE," Rear Admiral Robert W. Carius, Copy at the Jacksonville Historical Society Archives, Jacksonville University

Historical Reports of Commanding Officer, Naval Air Station Jacksonville, Copy at Historical Archives, Washington Navy Yard

Historical Reports of Commanding Officer, Naval Air Rework
    Facility & Naval Aviation Depot, Copy at Aviation History Division, Washington Navy Yard

Historical Reports of Commanding Officer, HS-1, HS-3, HS-5, HS-7, HS-9, HS-11, HS-15, HS-17, Copy at Aviation History Division, Washington Navy Yard

Historical Reports of Commanding Officer, VP-5, VP-7, VP-16, VP-18, VP-24, VP-30, VP-45, VP-49, VP-62, VW-4, VAP-62 Copy at Aviation History Division, Washington Navy Yard

Various Dates:

Florida Times-Union

Jacksonville Journal

JAX AIR NEWS

JAX FLYER

Naval Aviation News

Metropolis

REFERENCES CITED—"The Early Years"

Johnson, Robert E.
    1978   Archeological Investigations of 9CAM167 and 9CAM173 at King's Bay, Camden County, Georgia. Ms. on file U.S. Navy, King's Bay and the Department of Anthropology, University of Florida, Gainesville.

    1988   An Archeological Reconnaissance Survey of the St. Johns Bluff Area of Duval County, Florida. Ms. on file Florida Archeological Services, Inc. Jacksonville.

    1997   An Intensive Archeological Site Assessment Survey and Inventory of the Jacksonville Naval Air Station, Duval County, Florida. Ms. on file, U.S. Navy Facilities and Environmental Department. Jacksonville.

    1998a   A Phase II Archeological Investigation of Florida Inland Navigation District Tract DU-7 Greenfield Peninsula, Duval County, Florida. Ms. on file Division of Historical Resources, Florida Department of State. Tallahassee.

    1998b   Phase II Archeological Investigations of Sites 8DU5544 and 8DU5545 Queens Harbour Yacht and Country Club, Duval County, Florida. Ms. on file Division of Historical Resources, Florida Department of State. Tallahassee.

Russo, Michael
    1992   Chronologies and Cultures of the St. Marys Region of Northeast Florida and Southeast Georgia. *Florida Anthropologist* 45:107-126.

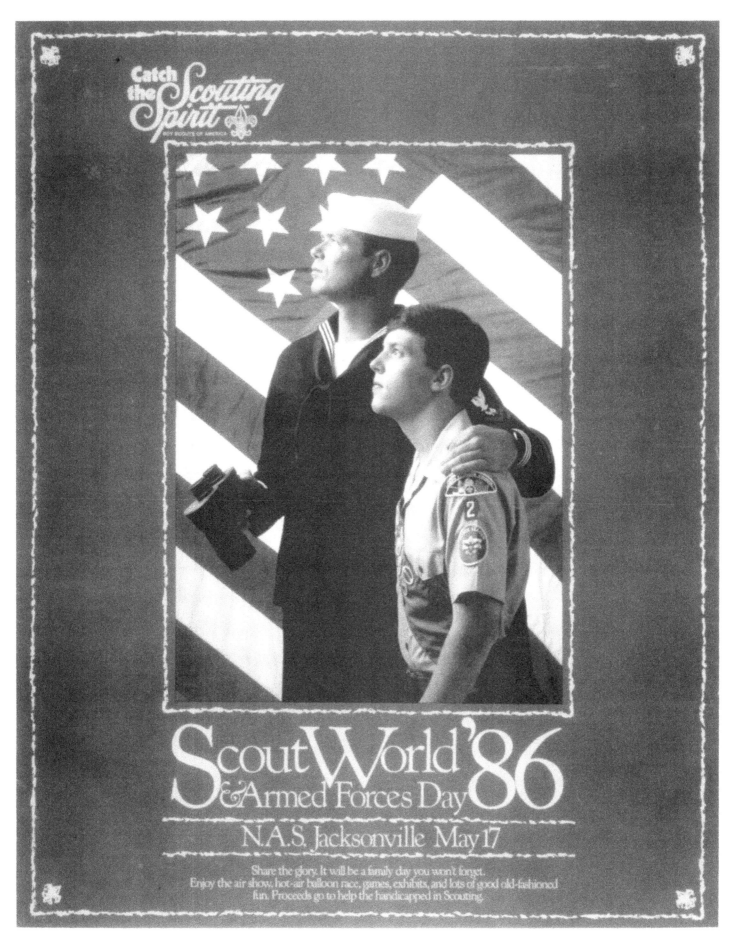

THANK YOU

to all those who made

this historic publication possible,

including

LOCKHEED MARTIN

whose ad is on the inside front cover

and

NORTHROP GRUMMAN

ORANGE PARK

The Place to Stay for a
Perfect Visit to NAS Jacksonville

300 Park Avenue North
Orange Park Florida 32073

1-800-533-1211

SCOUT WORLD 1991
AND ARMED FORCES DAY
N.A.S. JACKSONVILLE, MAY 18

BEHOLD. THE LEADERS OF TOMORROW.

Today's Scouts are tomorrow's leaders. Come and cross that bridge to the future at this year's Scout World and Armed Forces Day. Enjoy aircraft displays, games, exhibits and more. It's a day of old-fashioned fun for the entire family. Proceeds go toward programs to assist Scouts with disabilities.

SCOUTING A BRIDGE TO THE FUTURE

A I R S H O W 1 9 9 4

AIR SHOW '96
NAS JACKSONVILLE, FLORIDA
BIRTHPLACE OF THE BLUE ANGELS

OCTOBER 25, 26, & 27

And Saluting

1946 **FIFTY YEARS** **BLUE ANGELS** 1996

ALSO FEATURING

**NAVY, AIR FORCE & MARINE JET DEMONSTRATIONS
AND CIVILIAN AEROBATIC PERFORMERS**

NO PETS • NO COOLERS • FREE ADMISSION & PARKING • BRING THE ENTIRE FAMILY

GATES OPEN 8:30 A.M. – SHOW STARTS 11:00 A.M.

NAS JACKSONVILLE, FLORIDA IS 2 MILES NORTH OF I-295 ON HIGHWAY 17

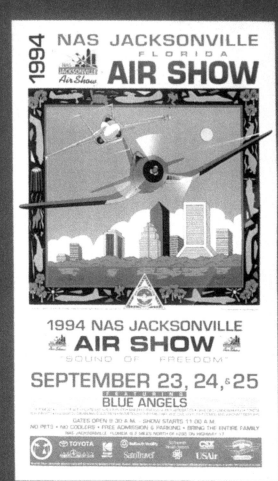

Printed in the USA
CPSIA information can be obtained
at www.ICGtesting.com
JSHW060052150824
68134JS00032B/2719